Amazon Echo Dot User Guide

Newbie to Expert in 1 Hour!

by Tom Edwards & Jenna Edwards

Copyright © 2016 by Tom Edwards & Jenna Edwards – All rights reserved.

AMAZON ECHO DOT is a trademark of Amazon Technologies Inc. All other copyrights & trademarks are the properties of their respective owners. Reproduction of any part of this eBook without the permission of the copyright owners is illegal – the only exception to this is the inclusion of brief quotes in a review of the work. Any request for permission should be directed to ReachMe@Lyntons.com.

Other Books By Tom & Jenna Edwards

Amazon Echo User Guide - Newbie to Expert in 1 Hour!

250+ Best Kindle Fire HDX and HD Apps for the New Kindle Fire Owner

All New 7" Fire User Guide - Newbie to Expert in 2 Hours!

Amazon Fire HD 6 User Guide - Newbie to Expert in 2 Hours!

Fire HD 8 & 10 User Guide - Newbie to Expert in 2 Hours!

Chromecast User Guide – Newbie to Expert in 1 Hour!

Amazon Fire TV Stick User Guide - Newbie to Expert in 1 Hour!

Amazon Fire TV User Guide - Newbie to Expert in 1 Hour!

Before We Start - Important!

Throughout this book we recommend certain webpages that might be useful to you as an Echo Dot owner and user.

A lot of the webpages we recommend are on Amazon, and a typical Amazon web address might look something like this:

www.amazon.com/dp/B00DBYBNEE?_encoding=UTF8&camp=1789&creative=9325&linkCode=pf4&linkId=UPTTB3CK67NSPERY&primeCampaignId=prime_assoc_ft

The link above takes you to the Amazon Prime sign-up page, but that's a lot of characters for you to type! So we've used what they call a Link Prettifier to make the links shorter and easier to use.

In this example the shortened link is - www.lyntons.com/USPrime - if you type that into your browser, it will save you lots of typing time, but still take you straight to Amazon Prime....

Contents

INTRODUCTION   1
   Welcome   1
   Do You Need This Book?   1
   How To Use This Book   2

1. WHAT IS AMAZON'S ECHO DOT?   5

2. SETTING UP YOUR AMAZON ECHO DOT   10
   What's in the Box   10
   Setting up the Main Unit   10
   Accessing the Alexa App   11
   Connecting to Wi-Fi   12

3. USING THE ALEXA REMOTE   14
   Pairing the Remote   14

4. USING THE ECHO DOT AS A BLUETOOTH SPEAKER   17
   Using Your Echo Dot Internal Speaker   17
   Pairing by Using the Alexa App   19
   Using Your Dot, a Wired Connection & An External Speaker   20
   Connecting Your Dot to an External Speaker Via Bluetooth   21

5. AMAZON ALEXA APP BASICS   23

6. MUSIC, RADIO AND AUDIO   26
   Amazon Prime Music: What it Offers   26
   Controlling Prime Music with your Echo Dot & App   **29**
   Pandora: What it Offers   31
   Spotify: What it Offers   34
   Setting Up your Spotify Account   34
   Controlling Spotify with the Alexa App   35
   Unlinking Spotify and Amazon Accounts   35

TuneIn: What it Offers	36
Setting Up your TuneIn Account	**37**
IheartRadio: What it Offers	38
Itunes on the Amazon Echo Dot	40
Audible: What it Offers	41
7. AMAZON ECHO DOT IN YOUR HOME	**43**
News, Weather, and Traffic	43
Movie Times	45
Find local businesses and restaurants	46
Timers and Alarms	47
Skills	48
To-Do and Shopping Lists	50
Sports	52
General Knowledge	53
Simon Says?	54
Control Your Home	54
Shopping	56
Amazon Echo Dot & Google Calendar	57
Add an Adult to Your Echo Dot Household	59
Voice Commands and Things to Try	60
8. THE ECHO DOT AND IFTHISTHENTHAT (IFTTT)	**61**
Using Your Echo Dot With IFTTT	62
The My Applets Page	66
Create Your Own Applets	67
9. TROUBLESHOOTING YOUR AMAZON ECHO DOT	**72**
The Echo Dot Doesn't Understand You	72
The Echo Dot Can't Give You an Answer	72
The Echo Dot Plays the Wrong Music	73
Voice Purchasing Code Doesn't Work	73
Amazon Echo Dot Remote Won't Pair	73
Echo Dot Won't Connect To Wi-Fi	74

Echo Dot Won't Connect to Bluetooth	75
Echo Dot Can't Discover a Smart Home Device	75
10. SECURITY	**78**
11. YOUR FUTURE WITH ALEXA	**81**
A Final Quick Reminder About Updates	82

Want the Latest Echo Dot News?

Before we start, we just want to remind you about the FREE updates for this book. The Amazon Echo Dot and indeed all media streaming services, like Apple TV, Roku and the Chromecast, are still in their infancy. The landscape is changing all the time with new services, apps and media suppliers appearing daily.

Staying on top of new developments is our job and if you sign up to our free monthly newsletter we will keep you abreast of news, tips and tricks for all your streaming media equipment.

If you want to take advantage of this, sign up for the updates here: www.lyntons.com/updates.

Don't worry; we hate spam as much as you do so we will never share your details with anyone.

INTRODUCTION

Welcome

Welcome and thank you for buying the **Amazon Echo Dot User Guide: Newbie to Expert in 1 Hour!**, a comprehensive introduction and companion guide to the exciting possibilities that the Dot bluetooth speaker and personal assistant has to offer.

Do You Need This Book?

We want to be clear from the very start - if you consider yourself tech savvy, e.g. the kind of electronics user that intuitively knows their way around any new device or is happy Goggling for answers, **then you probably don't need this book.**

We are comfortable admitting that you can probably find most of the information in this book somewhere on Amazon's help pages or on the Internet - if, that is, you are willing to spend the time to find it!

And that's the point... this Amazon Echo Dot book is a time saving manual primarily written for those new to streaming media devices, bluetooth devices and tech that works in tandem with your PC or mobile device.

If you were surprised or dismayed to find how little information comes in the box with your Echo Dot and prefer to have to hand, like so many users, a comprehensive, straightforward, step by step Amazon Echo Dot guide, to finding your way around your new device, **then this book is for you.**

Furthermore the Amazon Echo Dot is a brand new piece of kit and there will be new features, channels and games, not to mention Amazon Echo Dot tips and tricks, appearing constantly over the coming months. We will be updating this Amazon Echo Dot manual as these developments unfold, making it an invaluable resource for even the tech savvy.

Even if you are buying the first edition of this Amazon Echo Dot instruction book, never fear, you too can keep up to speed with all the new Amazon Echo Dot updates by signing up to our free email newsletter here - www.lyntons.com/updates - so you'll never miss a thing.

How To Use This Book

Feel free to dip in and out of different chapters, but we would suggest reading the whole book from start to finish to get a clear overview of all the information contained. We have purposely kept this book short, sweet and to the point so that you can consume it in an hour and get straight on with enjoying your Amazon Echo Dot.

This Amazon Echo Dot user manual aims to answer any questions you might have and offer Amazon Echo Dot information including:

- What is the Amazon Echo Dot and how does it work?
- What does the Echo Dot do?
- How to setup your Echo Dot
- How to setup Alexa
- How to manage your Amazon Echo Dot account
- Amazon Echo Dot and Alexa
- Amazon Echo Dot tips
- Amazon Echo Dot specifications
- Amazon Echo Dot settings
- And a general Amazon Echo Dot review

This Amazon Echo Dot tutorial will also look closely at Amazon Echo Dot features including:

- The Amazon Echo Dot voice remote (including Amazon Echo Dot voice commands)
- The Amazon Echo Dot Alexa app
- Amazon Echo Dot extras
- The Amazon Echo Dot shopping list and to do list
- Amazon Echo Dot radio, music and news
- How to use Amazon Echo Dot IFTTT
- Amazon Echo Dot smart devices (including home hubs and lights)
- Amazon Echo Dot accessories
- Plus much, much more....

And for further Amazon Echo Dot customer support we have links to

- Amazon Echo Dot customer service
- Amazon Echo Dot discussion forums
- Amazon Echo Dot feedback
- Amazon Echo Dot quick start guide
- Amazon Echo Dot videos
- Amazon Echo Dot Help Pages, help desk and community

As we will be updating this book on a regular basis we would love to get your feedback, so if there is a feature that you find confusing or something else that you feel we've missed then please let us know by emailing us at ReachMe@Lyntons.com. Thank you!

So without further ado let's begin.....

1. WHAT IS AMAZON'S ECHO DOT?

The Amazon Echo Dot is a puck sized personal assistant that is becoming more useful all the time. Amazon is consistently developing new capabilities and updating your Echo Dot via its Wi-Fi connection. While you might first think that Amazon Echo Dot is little more than a technological novelty that can do a few cool things, it has become a valued asset in the time we've been using it.

We turn to the Echo Dot throughout the day for news, weather, music and general information. It assists us in many ways with time management and scheduling, home automation, online purchasing, communication and social media, with more uses being added on a regular basis.

The Echo Dot is the smallest of three similar devices. The original Amazon Echo is just like the Echo Dot but with a large internal speaker that gives much better audio playback, it needs to be plugged into the mains. The Amazon Tap is a portable version of the Amazon Echo which you can take anywhere. The idea of the Echo Dot was to give you the functionality of the original Echo in everyroom of your house.

This guide will help you get the most fun and functionality from your Amazon Echo Dot. We want to set your mind at ease from the very beginning: **Setting up the Amazon Echo Dot is simple**. From there, each step in tailoring this smart device to fit your lifestyle is easy and takes just a few minutes.

To personalize the interaction, Amazon has given the device the name "Alexa," and that is the wake word used to activate it.

Talking with the device really is like a conversation due to its pleasant and fluid voice, and "she" will soon be assisting you in lots of wonderful ways.

By the way, if you find it odd to address an inanimate object using a woman's name, the alternate wake word is "Amazon." We've tried both and have finally stuck with Alexa. It seems perfectly natural.

The Alexa App is a vital piece of the Amazon Echo Dot system. In short, the App is a piece of software easily downloaded to your smart phone, tablet or PC, and it works in tandem with the Echo Dot to allow you to get the most from its capabilities.

The app is, essentially, the remote control center for the Echo Dot. We explain it fully in Chapter 5 of this guide.

Voice recognition software is the key to your communication with the Amazon Echo Dot. The device is programmed to understand North American English and will comprehend most users without a problem.

We've had some fun speaking to the device using exaggerated intonations and inflections including a variety of badly performed accents, and Alexa has understood us remarkably well. If you do, however, experience a communication barrier, don't despair.

The Amazon Echo Dot App provides Alexa with voice training. The technology "teaches" the Echo Dot to understand you when you say such things as "Alexa…"

- What time is it?
- Set an alarm for 8 a.m.
- Set the timer for 10 minutes.
- How is traffic?
- How many cups in a quart?
- Tell me a joke.
- Play music by Bruno Mars.
- What is the weather in Chicago?
- Add olive oil to my shopping list.
- Remind me to do the laundry.

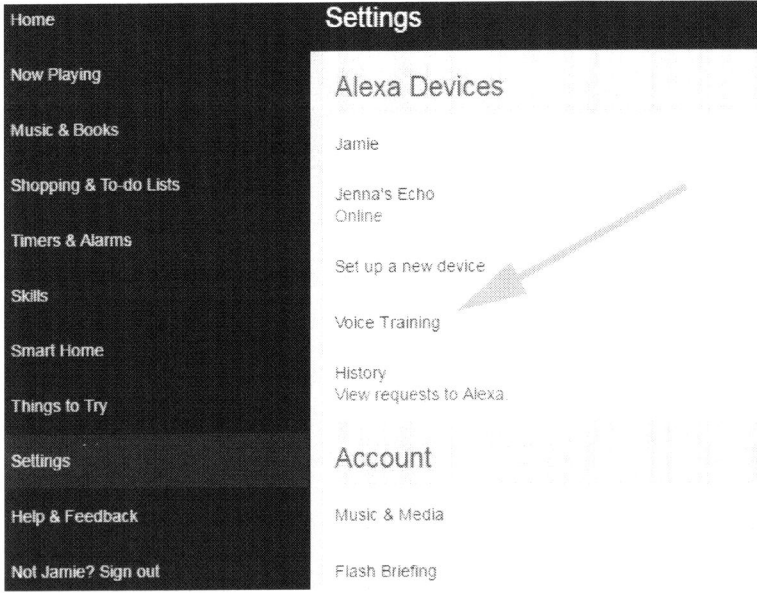

The Voice Training tab is found within **Settings**.

Voice Training

Introduction

Voice training can help Alexa work better by learning how you speak. In this short session there will be 25 phrases to read aloud. Remember to:

1. Speak to Alexa as you normally would
2. Speak from the same distance you normally would
3. Don't use the Alexa device remote

Start Session

Cancel Voice Training

Now, let's introduce this attractive cylinder and how it works. You'll notice four buttons on top of the Echo Dot. You have 2 volume buttons, volume up and down, and two others. The light ring will respond with a white light to indicate the volume level of the Echo Dot. You can also give voice commands for this such as, "Alexa, louder volume"

The third button is the Microphone On/Off Button.

When they are switched on, the microphones are listening for you to speak the wake word. When turned off, the mics can't hear you. It's that simple.

You might turn off the microphones if you're using the Amazon Echo Dot remote, which is sold separately. We'll explain how easy it is to communicate with the Echo Dot using the remote in Chapter 3.

The final button on top, the one with the single raised dot, is called the **Action Button**. It can be used to wake the Echo Dot in lieu of saying the wake word, turn off a timer or alarm sound and enter Wi-Fi setup mode.

A **light ring** circles the top of the device and serves as an indicator of the Echo Dot's status. Here's a complete list of the light ring colors and what they mean:

- **Solid blue with spinning cyan lights:** The Echo Dot is starting up when first plugged in.

- **All lights off:** The Amazon Echo Dot is waiting for you to wake it using the wake word or the action button, or it is unplugged.

- **Solid blue light except for a small cyan section:** The Echo Dot is listening to you or replying; the cyan shows which of the seven microphones is picking up your voice.

- **Flashing blue and cyan lights:** The Amazon Echo Dot is processing your request.

- **Alternating blue and cyan lights:** The device is giving a longer answer.

- **Solid red light:** The microphones on the Echo Dot have been turned off - To turn them on, press the Microphone button on top of the unit.

- **White light:** The Echo Dot's volume is being adjusted.

- **Oscillating violet light:** An error has occurred during Wi-Fi setup.

The Amazon Echo Dot is equipped internally with seven microphones to pick up your voice from any direction. They are turned on using the Microphone button. When it is on, the device starts listening and is ready to assist you when you say the wake word or press the action button.

2. SETTING UP YOUR AMAZON ECHO DOT

The setup was super-quick, and much of it happened automatically with little or no input from us. Here are the details.

What's in the Box

- Amazon Echo Dot
- Micro-USB cable
- Power Adapter
- Quick Start Guide

The Alexa Voice Remote for the Echo Dot, which you might find very handy, is sold separately by Amazon for $29.99. However, it is not needed in order to use your Echo Dot.

Setting up the Main Unit

Locate it where you spend most of your time. Much of the time, our Amazon Echo Dot speaker is in the kitchen on the breakfast bar, and our Echo Dot frequently moves between rooms, the bedroom, living room and office. When it is plugged into the new location, it reconnects to our Wi-Fi system by itself in 30 to 75 seconds. Of course at just $49,99 Amazon have priced the Dot in the hope that you will buy several of them and scatter them around the house as needed.

Amazon recommends that the device be placed at least eight inches away from walls and windows to ensure clear communication, perhaps to avoid confusing the Echo Dot with an Echo Dot! We've tried it against a wall, and it worked fine. Just keep that tip in the back of your mind in case your Amazon Echo Dot parked on a window sill or near a wall is having communication problems.

Simple to Set Up & Use

1. Plug in Amazon Echo Dot
2. Connect to the Internet with the Alexa App
3. Just ask for music, weather, news and more

Plug the power adapter into the Echo Dot and into a 110-120 volt outlet. The light ring will turn blue to acknowledge that it is powered and then turn orange. At this point, Alexa will greet you with "Hello".

Now, it's time to get familiar with the Amazon Echo Dot App.

Accessing the Alexa App

Setting up the Amazon Echo Dot takes just a few minutes using the quick-start guide included in the package, and it worked flawlessly when we tried it. First up visit https://alexa.amazon.com on the device (mobile phone, tablet or pc) that you intend to use for controlling the Echo Dot via the Alexa app. Alexa is your voice activated assistant so you need the Alexa app to control your Echo Dot.

When you click Download now you will be redirected to the relevant appstore for your device where you can download the app. Should you for some reason want to do this independently of the setup process you can visit this page - www.lyntons.com/echoapp - to access the following app stores:

- Amazon Appstore

- Google Play

- Apple App Store

The app can also be used on your computer using a supported browser, currently Chrome, Safari, Firefox and Internet Explorer. Again just type in https://alexa.amazon.com to get started. When using your computer to access and operate the app everything is done in your browser so you won't be downloading any software using this method.

Connecting to Wi-Fi

The Alexa App walks you through the short process of connecting the Amazon Echo Dot to a Wi-Fi network. Have your Wi-Fi password available. The password is usually located on the Wi-Fi router. If it's not there, contact the company that provides your Internet service.

Note: the Echo Dot does not support enterprise or ad-hoc/peer-to-peer networks, though it is unlikely that your network is one of these types.

Once the app is downloaded, go to the Amazon Alexa app homepage. Open the navigation panel on the left, and select **Settings > Echo Dot > Set up a new Echo Dot**. By the way, this sequence means you should:

1. Select **Settings** from the options on the Amazon Alexa App homepage

2. Select **Echo Dot** from the options that appear

3. Select **Set up a new Echo Dot** from the fresh set of options you're given

The next step is to press and hold the **Echo Dot Action Button** for about five seconds. As you might recall, that is the button on top of the cylinder with the single raised dot. The light ring will turn orange while your mobile device, if using one with the Echo Dot, connects to it.

In the app, a list of one or more Wi-Fi networks will appear. Choose your Wi-Fi network, and enter the network password if necessary. Select **Connect**. If your network isn't on the list, scroll down to select **Re-scan** to search again or choose **Add a Network** and follow the instructions that appear.

When we did this, we didn't strictly time the process. It seemed to take more than a minute and less than three for the Echo Dot to connect to the network. When it did connect, a confirmation message appeared on the **Set up a new Echo Dot page** of the app....so be patient.

You're now ready to return to the app Home Page, which you'll find at the top left of the screen, to begin using your Amazon Echo Dot.

Start by addressing the Echo Dot with the default wake word "Alexa". The wake word can be changed to "Amazon" by selecting **Settings > [Your Name]'s Echo Dot > Wake Word > Amazon > Save**.

Where it says [Your Name]'s Echo Dot above, we mean the name you change your Echo Dot to during the "Set up a new device" process. We named our Echo Dot 'Jenna's Echo Dot', so in our case it would be **Settings > Jenna's Echo Dot > Wake Word > Amazon > Save**.

If we ever want to change the name there's an option to change it, just go to **Settings > Jenna's Echo Dot > Device name > Edit**

3. USING THE ALEXA REMOTE

If you've purchased an Amazon Alexa voice remote, now is a good time to set it up and put it into action. While the remote isn't needed, those who have one find it to be very useful.

Pairing the Remote

To start, make sure the two AAA batteries have been inserted and that they're oriented properly. In most cases, the remote will automatically pair with the Echo Dot. If it does not:

- Go to the Alexa App
- Open the navigation panel on the left
- In the Settings, select your Echo Dot
- Then select **Pair Remote**

The Amazon Echo Dot will search for the remote and connect to it shortly. If this doesn't work, press and hold the **Play/Pause** button on the remote for a few seconds, and repeat the process.

If you ever want to remove the remote and pair a new remote to the Echo Dot, choose the **Forget Remote** option in the settings under your Echo Dot. Add the new remote using the steps above.

One of the remote's most popular features is that it gives you the ability to use the Amazon Echo Dot even when the microphones have been turned off. The Echo Dot is still able to receive questions and commands via the remote control and will respond. Use the button on top of the cylinder to turn off the mics. The light turns red and stays red while the microphones are off.

We use this feature in two different scenarios. First, we use the remote when there is noise in the room - people conversing nearby, the TV or sound system on - that might disrupt communication.

Imagine starting with, "Alexa...", and as you're asking for weather information, the microphones pick up a football game on TV. You'll get a response such as, "Hmmm, I can't find the answer to the question I heard."

When using the remote, the Echo Dot receives requests digitally rather than audibly, and it processes them without audio interference.

We also choose this feature at a distance from the Echo Dot and don't want to have to speak loudly to converse with it (or are too comfortable on the couch to get up!) Here are tips for such occasions:

- Keep the remote handy, so you won't have to get up from your relaxing spot.

- You might need to turn up the Echo Dot's volume by requesting, "Alexa, volume 8," for example, or, "Alexa, louder," and then making your request again.

- Communicate using the remote the same way you would speak directly to the Echo Dot. There's no need to yell, like we did at first, because it is the remote that hears your request and sends it on to Alexa.

To communicate via the remote, simply speak into it as you would speak directly to the Echo Dot. If there is consistent noise near the Echo Dot, you might need to turn off the microphones via the button on top of the cylinder. The light ring turns red and stays red when the mics are off. Push the button again to turn them back on.

If you already have the Amazon Fire TV, this remote should be quite familiar. If not, learning to use it will still be very easy.

Please note: You can only connect the Echo Dot to one Bluetooth device at a time, so for example you can't connect your Echo Dot to the Alexa voice remote if it is currently connected to a Bluetooth speaker. That makes the Alexa voice remote a slightly less appealing purchase for those with an Echo Dot.

4. USING THE ECHO DOT AS A BLUETOOTH SPEAKER

As you probably already realize, the Echo Dot is a paired down version of the Amazon Echo. While the Echo Dot does contain a speaker it is nowhere as sophisticated or as powerful as the speaker on the larger Amazon Echo. That's not to say the Echo Dot's speaker is useless, we've found that it's perfectly good enough for everything except high quality music playback.

So if you're getting spoken information from Alexa (news/traffic/general knowledge), listening to an audio book or simply want to have some quiet music playing in the background then the Dot's small internal speaker is sufficient.

But if you want your Echo Dot to really sing then you can connect it to a separate external speaker, either via Bluetooth or a wired connection, and your sound quality will be just as good, if not better depending on your external speaker, than using the original Amazon Echo.

So far so good!

However the way you use your Dot will affect some of the things that you can do and how you should set things up. So let's now look at three different sound set up scenarios and see which is best for you.

Using Your Echo Dot Internal Speaker

As you work your way through this guide you will discover all the different features that the Echo Dot offers. There are lots of them and they are the main reason you bought your Dot in the first place!

That said you can also use your Dot simply as a standalone Bluetooth speaker, all be it a not very powerful one. But why would you want to do this? Well the simple answer is that we all have mobile phones or tablets or computers that we spend a lot of time on and these devices also have lots of audio content, like music, podcasts, audiobooks.

We listen to a lot of podcasts and we do that almost exclusively through our mobile phones. Therefore for us it's extremely useful, when we're at home, to stream these podcasts through an external speaker…an external speaker like the one in the Echo Dot!

So, if that's something you would like to do this is how you go about it:

To stream content from any other Bletooth enabled device you may have using the Echo Dot as your speaker, it must first be paired to the device containing the content you want to hear. This is done by finding and accessing the Bluetooth settings of the PC, phone or tablet you want to pair your Echo Dot with.

Obviously the location of the Bluetooth settings will vary from device to device, they are typically found in the general Settings menu of your device. On a computer, they are usually found using the search function. For example, you can use the search box on the Start panel for Windows to find the Bluetooth settings.

Here's a quick guide to pairing the Echo Dot with a mobile device using either voice commands or the Amazon Alexa app.

Say, "***Alexa, pair*****"*. The Echo Dot will reply with "Ready to pair" and will give you instructions on how to proceed doing the following:

- Browse the Bluetooth settings on the device to locate and choose "Echo Dot-###".

- Open the Bluetooth settings, and the Echo Dot should appear within a few seconds. When it does, select it.

- When the pairing is completed, the Echo Dot will say, "Connected with Bluetooth".

- When you no longer want the mobile device paired to the

AMAZON ECHO DOT USER GUIDE

Echo Dot, say, "Alexa, disconnect" and the pairing will be terminated.

Pairing by Using the Alexa App

• Open the left navigation menu, and select **Settings**

• Select the Echo Dot. If you have more than one Echo Dot, choose the one you want to pair by selecting the name you've given it.

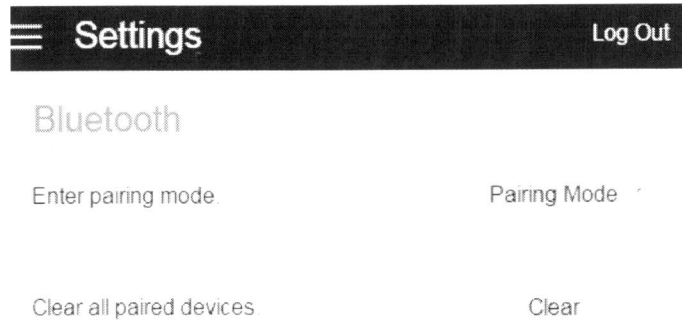

• Select **Bluetooth > Pairing Mode**, and the Echo Dot will say, "Ready to pair".

• From the Bluetooth Settings on your mobile device (typically accessed from the general Settings menu), select "**Echo Dot###**".

• Open the Bluetooth settings on the device, and choose the Amazon Echo Dot you want to stream content to. The Echo Dot should appear within five seconds.

• When pairing is completed, the Amazon Echo Dot will reply, "Connected with Bluetooth".

• When you're finished using the Echo Dot as a Bluetooth speaker, say, "**Alexa, disconnect**" and the connection to the mobile device will be ended.

When a Bluetooth device is paired to your Echo Dot, you'll enjoy hands-free control with commands such as, *"Alexa...*

- ***Play***
- Pause
- Previous
- Next
- Stop
- Resume
- Restart

It should be stressed that before you can use these commands you need to start playing something via your paired device, play an album in your iTunes app for example or start playing and audiobook or podcast.

We found through experience, though the information is on Amazon too, that if you request specific songs, albums or artists from the music stored on your phone or other Bluetooth device, the Echo Dot will disconnect from the device and search for the music in your Amazon Echo Dot library. So you can only use these voice commands to control content that is already playing.

To reconnect to the Bluetooth device, simply say, *"Alexa, connect"*, and press play on the device to continue listening.

Unfortunately, the Echo Dot can't yet receive phone calls, text messages or other notifications from your mobile device. Additionally, audio from the Amazon Echo Dot can't be sent to other Bluetooth speakers or headphones.

Using Your Echo Dot With a Wired Connection To An External Speaker

For this scenario let us now imagine that you want exactly the same set up, as we've just discussed in the previous scenario, but this time you want excellent audio quality. You want all the features of the Echo Dot **plus** you want to be able to stream audio content from or phone, tablet or pc and you want that audio to sound great.

In this case you follow exactly the same instructions as above and pair your phone or pc with the Dot, but you also need to connect your Echo Dot to an external speaker via a 3.5mm audio cable. You will need to buy this audio cable separately.

Just so you know, when your Echo Dot is connected to a speaker via the audio cable the external speaker will override (shut down) the internal speaker of your Dot... even when the external speaker is switched off. So when the two are connected via an audio cable make sure the external speaker is turned on before trying to interact with your Dot.

Connecting Your Echo Dot to an External Speaker Via Bluetooth

If you want the convenience of connecting your Dot wirelessly to a Bluetooth enabled speaker then it's pretty simple. Yet again follow the pairing instructions we gave you earlier. The only difference will be that to turn on Bluetooth connectivity on your external speaker you will need to press a button on the speaker. Where that button is will depend on the Bluetooth speaker your using.

Connecting wirelessly means you can place your speaker anywhere in the room, it doesn't have to be near/wired up to your Echo Dot.

However the downside is that you can no longer stream content from any other device, like a mobile or pc, while using your Echo Dot. Basically your Echo Dot can only pair with one Bluetooth device at a time so if it's paired with your speaker it can't then pair with your phone at the same time.

If you are someone who is really invested in Amazon and all their technology this may well not be a problem. If you already listen to all your music using the Amazon Music Library or if you're happy to listen to your podcasts via TuneIn which is supported by the Alexa App (more on that later) then you won't care about streaming audio from your phone because you will already have access to all the content that you like via your Dot and the Alexa app.

So let's now move on and look more closely at the Alexa app, what it can do and what content and great features you can access though it.

5. AMAZON ALEXA APP BASICS

The smart Echo Dot and the Amazon Alexa app work together as a team to create a really cool personal assistant that we expect to become more competent as time goes by, via Amazon updates and the addition of new third-party partners. Together, the app and device form a system called Amazon Echo Dot. This chapter is an overview of the Alexa app and how to use it.

We first explored the app on our laptop because navigation is simpler and the large screen is easier to see. Once we were familiar with the app, working with it from a cell phone is a snap.

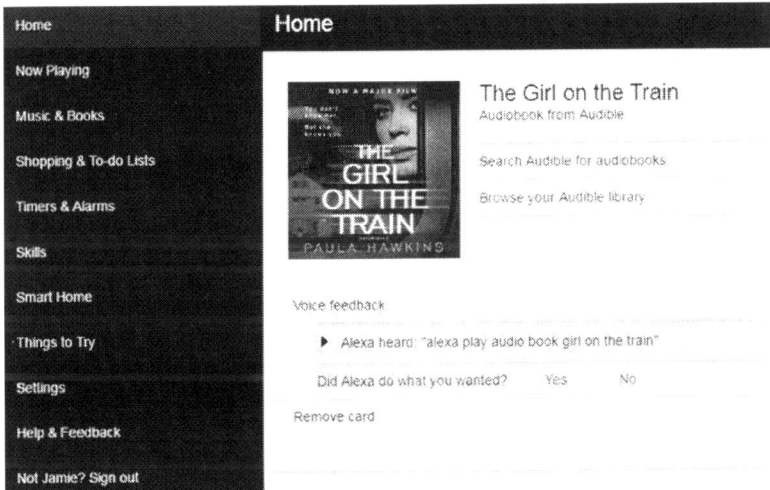

The Home Screen is divided into two sections. The Menu is on the left edge, and the rest of the screen is taken up with the dialog cards. One card is created for each request made of Amazon Echo Dot.

When you select the down arrow on any card, the option is given to provide feedback to improve your Amazon Echo Dot's voice recognition. What Alexa heard is listed with the question, "Did Echo Dot hear you correctly?" Answer ***Yes*** or ***No***. You've also got the option to remove the card.

At first, we gave feedback and deleted a few cards to reduce clutter, but we mostly forgot about it pretty quickly.

Now the only time we choose the down arrow and give feedback is if the Echo Dot did not hear us correctly, a rare occurrence. We also occasionally scroll back through the cards to locate a song/playlist/station we enjoyed but forgot to get the name of.

Depending on the type of request that produced the card, you'll have more options. For example, if you've requested Echo Dot to play something from your Amazon Music Library, the card includes options to **Search the Amazon Digital Music Store** for similar music, Search **Amazon Prime** for free music and **Browse your library** for similar choices.

Cards for information requests will include the option to Search Bing for the topic.

If you choose the **Learn More** option in blue on any card or at the top of the list of cards, the Amazon Echo Dot FAQs appear. There are 12 questions, and they're worth reading over.

The Amazon Alexa app menu on the left gives these choices:

- **Home:** When Home is selected, your list of cards appears

- **Now Playing:** Shows you in detail exactly what you are currently listening to

- **Music & Books:** This is where you can find info on your Amazon Music Library, Amazon Music subscription, Spotify, Pandora, iHeartRadio, TuneIn and Audible.

- **Shopping & To-Do Lists:** Create shopping or To-do lists manually or by voice

- **Timers & Alarms:** Set, pause, continue and cancel timers up to 23:59:59 and Set an alarm for up to 24 hours, and toggle it on or off.

- **Skills:** Discover all the third part apps that add further functions and features to your Echo Dot

- **Smart Home:** This is where you connect your smart home devices to your Echo Dot

- **Things to Try:** This list shows you how to get the most from Amazon Echo Dot through topics like What's new, Do more with your Echo Dot and tips for managing the app.

- **Settings:** We'll refer to Settings throughout this guide to help you manage your Amazon Echo Dot and account.

- **Help & Feedback**

You can manage the history of your dialog with Amazon Echo Dot in much the same way as you manage your computer's history. Go to ***Settings > History*** where you'll find the complete list of requests and commands.

We occasionally scroll down the list to find a station we were playing or retrieve an answer to a question we asked. Provide feedback or delete any request by selecting the down arrow.

If you prefer to delete all voice recordings, go to ***Manage your Content and Devices***, and select the ***Your Devices*** tab. Choose the Echo Dot, and a popup window will appear where you can delete or cancel the delete request.

Keep in mind that Amazon Echo Dot has been learning how you speak to give more consistently accurate answers. If you delete your voice recordings, you might find that the Echo Dot's ability to understand you is reduced. See our **chapter on Security** for more information.

Now that you've been introduced to your new personal assistant, it's time to get to know Alexa more thoroughly and find out what her true capabilities are.

We've created this section from our experience, and it's intended to be a guide you can refer to in the days ahead to maximize the fun and usefulness of your Amazon Echo Dot and to minimize the hassle. Let's start with the Echo Dot's strongest feature...Audio!

6. MUSIC, RADIO AND AUDIO

We're music fans, and you probably are too. Amazon gets this, so it built the Echo Dot to deliver real listening pleasure. Let's look at the top music services supported by the Amazon Echo Dot.

Amazon Prime Music: What it Offers

There's a long list of benefits for using Amazon Prime including Prime Music. There's a 30-day free trial if you want to try it out before making a commitment.

With Prime Music, available to Prime members in the U.S. and UK, you have unlimited ad-free access to more than two million songs in addition to complete albums and stations created for all major music genres. If you're not interested in Amazon Prime then rest assured there are plenty of other free music options available that we discuss later in this chapter.

Within the Amazon Alexa app click on the **Music & Books** tab, you will then see two Amazon music tabs to choose from. The first tab is **My Music Library** which contains all songs, albums and lists you have bought and downloaded in your standard Amazon account.

Using the Amazon Music Importer, you can add up to 250 songs from your computer to your Amazon Music Library to play on Amazon Echo Dot and other devices.

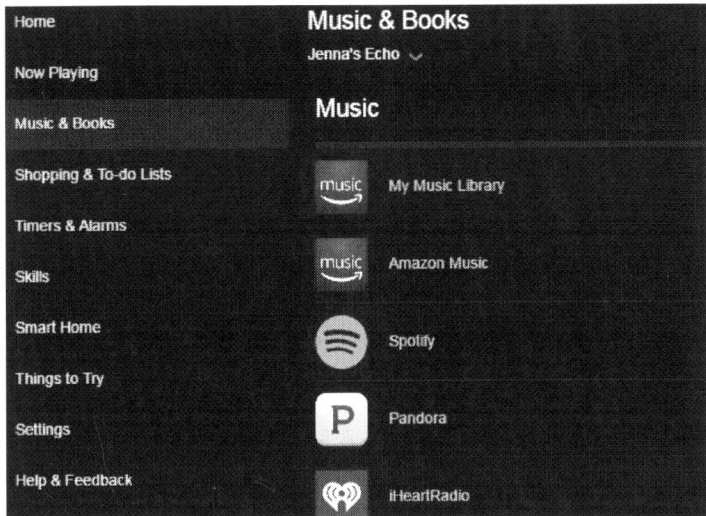

We explain more about Prime Music and downloading or uploading music to and from your Amazon Music Library in our book **Amazon Prime & Kindle Unlimited Newbie to Expert in 1 Hour!**

The second tab is the Amazon Music tab where you can browse Prime Music Stations, Genres and Playlists that are available to Prime members. Amazon are now offering a further music subscription called Amazon Music Unlimited, for a fee of $3.99 a month you can listen to even more music via your Echo Dot. Personally we find that the music options offered via our Prime membership is suffecent for our needs.

Setting Up your Prime Music Account

When you order Amazon Prime, you'll be taken to the Prime Welcome page where you can explore its benefits including Amazon Prime Music.

In addition to the Echo Dot, you can listen to Prime Music on a wide range of devices including iOS and Android phones, your PC or Mac and Amazon products such as the Fire phone, Fire tablet and Fire TV Stick.

It would be an understatement to say that there are a lot of musical options on the Prime Music homepage. Frankly, there are so many options, and so many of them overlap, that it was initially confusing.

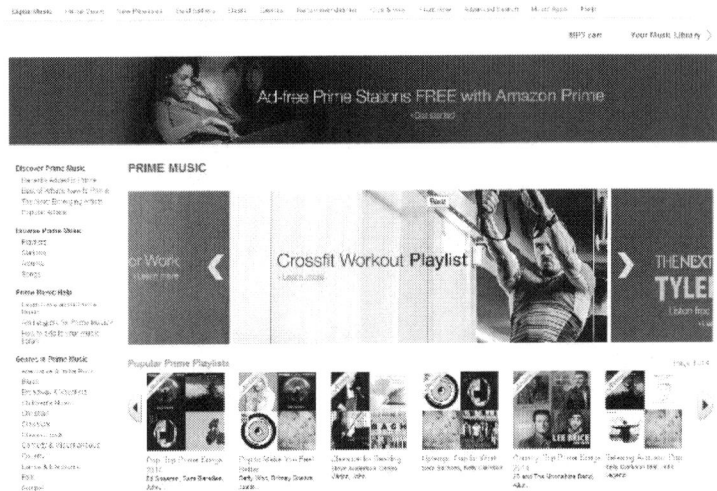

Think of it this way: The Amazon Prime music categories discussed below attempt to organize the entire catalog of modern music into logical groups which Amazon term as Stations, Genres and Playlists. There aren't hard and fast distinctions between these groups.

We quickly realized that there are just too many options to get a handle on easily so we adopted an adventurous, open-minded attitude. We pick a station or playlist that sounds good to us, and wait to hear what songs are included in the collection. We then note the station and return to it or avoid it based on our tastes.

Okay, with that perspective in mind, here is an overview of the musical groupings. Take a relaxed hour or so to explore what's available and get a feel for what directions you want to explore first.

All Stations: The dozens of stations cover all the musical styles we're familiar with, and we're music buffs, plus lots we're not acquainted with but look forward to exploring. Here's something we didn't expect when we began to listen: Stations named for a specific artist play that artist as well as similar artists. For example, the James Taylor station plays plenty of JT, of course, but also Christopher Cross, Joni Mitchell, Jim Croce, Gordon Lightfoot, Carole King, Jackson Browne, etc.

Genres: The genres are divided into favorites such as Alternative, Blues, Children's Music, Classical, Classic Rock, Decades, Gospel and

Pop. Each genre offers stations named for artists, decades and styles.

Playlists: There are thousands of Playlists that give you a more specific slice of the types of music you like, and your options are incredibly diverse. You can search Playlists by genre, artist, mood or decade to locate those that appeal to you. Select "Add Playlist to Library" and the list will be stored there to be played on your Amazon Echo Dot, phone or other device using the Alexa app.

Albums & Songs: More than 1.5 million songs and almost a thousand complete albums are available for you to add to your Music Library.

See what we mean? Amazon Prime offers a torrent of music, and there are innumerable places to jump into the rushing waters. We think you'll enjoy the ride, as we do.

Controlling Prime Music with your Echo Dot and App

Using voice, it is as easy as saying, "***Alexa, play Classical for Focus***" to get the music started. The Amazon Echo Dot will reply with "The Playlist Classical for Focus," and Bach's Brandenburg Concerto No. 3, Handel's Water Music Suite No. 1 and other great pieces will fill the air through the Echo Dot's quality speaker. Once the chosen list is playing, Alexa will respond to commands that control the music. These include, "***Alexa:***

- *What's playing?*
- *Turn it up or Softer*
- *Volume 5*
- *Skip or Next Song*
- *Pause, Resume or Continue*
- *Loop*

- *Stop*
- *Continue*
- *Buy this song*
- *Add this song* (to your Amazon Music Library)

When using the Alexa app instead or your voice, simple taps or clicks are all that are needed to choose the music you want to play. When you've made your choice, a player appears with options for play, pause, skip ahead, go back or shuffle the music.

You can also select the Thumbs Up or Thumbs Down symbols, or say, "Alexa, I like/dislike that song". Alexa will respond with, "Okay, rating saved", and Prime will remember your preferences for the next time you select that Station or Playlist.

Pandora: What it Offers

Similar to Tunein and iHeartRadio, Pandora offers hundreds of customized stations in more than 40 genres. There are standards such as Blues, Classical and Comedy and quite a few unique offerings like BBQ, Festivals and Musica Romantica.

Setting Up your Pandora Account

Select the Pandora tab from within **Music & Books** App to get started. You'll have the option of signing in to an existing Pandora account or registering a new one. A new account requires providing your email, user name, password, gender and zip code.

Once you've signed in or created an account, Pandora is immediately supported by Amazon Echo Dot. For those with an existing account, your stations will appear in My Stations, and you can alter or delete them as you wish.

Pandora Basic is free, but short commercials occasionally play between tracks. An upgrade to a premium account is available on the Pandora site. Premium accounts are ad free and allow more song skips. Currently, the upgrade is $4.99 per month or $54.89 annually.

You can use this link to upgrade - www.pandora.com/one

Controlling Pandora with your Echo Dot and App

If you've created a new Pandora account, start by selecting the **+Create Station** box at the top of the Pandora page on the Echo Dot App. From there, you can use the box to search for an artist, genre or track, or you can see what's already available in the Browse Genre section.

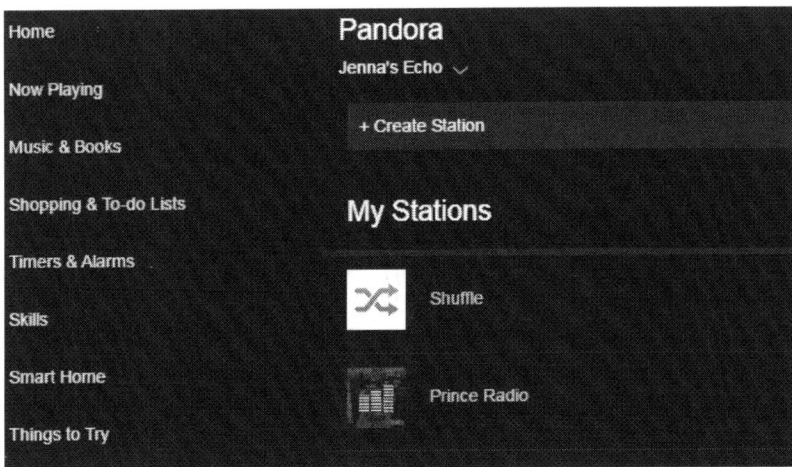

When you select an existing station, it begins to play and also appears in the list entitled My Stations on the Pandora Home Page.

Select the down arrow to have the option of deleting the station. Pandora allows you to create up to 100 of your own stations to supplement what is already available. For example, we're fans of a musical genre called reggaeton, and there's not a pre-existing station for it.

However, when we typed reggaeton into the search box, several tracks came up. We selected the track Reggaeton Latino, and a station was created with that tune and similar songs. We spell this out in detail because there are no "how-to's" given on the app. Some tasks, such as creating our own stations, we learned how to do through trial and error. However, as far as stations are concerned, there are already so many customizable stations within each genre of music, creating your own might not be something you'll want to do anyway.

Any station's playlist can be customized by eliminating unwanted songs from the list. Do this by saying, "***Alexa, thumbs down***", or by selecting the downward-pointing thumb on the player at the bottom of the screen.

Note, however, you'll only be able to delete or skip three songs on any list. The fourth attempt will produce a drop-down menu saying, "Our music licenses limit the number of songs you may skip." In other words, you'll have to listen to the remaining songs all the way through or choose a new station to play.

The Pandora player offers options to play, pause, go forward and go back. If you click or tap the small icon of the album cover for the song playing, you'll get the full page view with a larger icon and a list of the songs as they play.

You'll also have the option here to look at your stations history and the queue of songs that have played on the current station. Choose the down arrow of any of the songs in the queue to rate it, create a song or artist station, shop for the music at the Amazon Digital Store, Bookmark the song or give the song a rest by selecting "***I'm tired of this track***".

You can use voice commands with Pandora on Echo Dot, just as you can with the other services.

For more voice command options for Pandora, IHeartRadio, TuneIn and Prime Stations visit this page on Amazon - www.lyntons.com/PandoraHelp.

Spotify: What it Offers

This Swedish streaming service hooked up with Alexa after iHeartRadio and several others were onboard. In similar fashion, Spotify delivers music (30 million songs and counting) and podcasts to Echo Dot users through searchable artists, albums, labels, genres and playlists.

While Spotify has both Free and Premium subscriptions, you'll need to go with Premium to listen to the service with your Echo Dot. The current cost for Spotify Premium is $9.99, but a 30-day free trial is available that gives you ad-free streaming, unlimited skips and play any track features.

Setting Up your Spotify Account

Once you've established a Spotify accounting, synching it to the Alexa App on your laptop or phone will take just a couple of minutes.

- Open the Alexa App

- Browse the Menu, and select Settings

- Locate Music & Media, and choose Spotify

- Choose Link Account

- Log into Spotify with Facebook or with your username and password

Controlling Spotify with the Alexa App

Using voice commands: Speaking directly to the Echo Dot or using the Echo Dot remote, ask, "*Amazon/Alexa, play Twenty One Pilots on Spotify.*" It's important to include "*on Spotify*" or Alexa will typically select music from Amazon Prime or another music service first. You can ask for a specific song name, playlist name, genre, artist, composer or Discover Weekly to hear what's new. For example, say, "*Amazon/Alexa, play songs by Carrie Underwood on Spotify*," or "*Amazon/Alexa, play Work by Rihanna on Spotify.*"

Command options are what you'd expect: Pause, resume, stop, mute, previous, next, shuffle, skip this song, volume 1-10, volume up/down, etc. Remember to use your wake word.

You can get answers to questions like, "*what song is playing?*" and "*who is this artist?*"

Using the Alexa App: The Player in the App shows you the current artists and gives you standard command options such as play, pause, skip, previous, next and shuffle.

Using the Spotify App on your phone: Select Amazon Echo Dot from the device list. From there, voice commands will work.

Unlinking Spotify and Amazon Accounts

If you don't want the accounts linked:

- Open the Alexa App
- Browse the Menu, and choose Settings
- Choose Music & Media
- Choose Spotify from the list
- Choose Unlink account from Alexa
- Choose the Unlink option

TuneIn: What it Offers

TuneIn is completely free, and setting up an account isn't necessary. However, having an account allows you to "Follow" your favorite stations and shows and to Share them on Facebook, Twitter and Google Plus and Tumblr.

On TuneIn, you've got the opportunity to access more than 100,000 Internet radio stations including FM, AM, HD, LP and digital. Brands featured include ESPN, NPR, Public Radio International, CBS and C-Span. Your browsing options include much more than Music.

There is Local Radio, Sports, News, Talk and By Location along with popular shows featured in all of those genres. There are more than 4 million podcasts, concerts and interviews available too, a fact that sets TuneIn apart from its competitors.

Setting Up your TuneIn Account

As mentioned there's no need to set anything up to use TuneIn, you can access and browse everything on offer by following the instructions below.

However if you do want to take advantage of some further features you can join TuneIn at tunein.com. Signing up is a brief procedure that includes choosing a user name with an associated email address and a password. You also have the option of signing in with an existing Facebook or Google+ account, a step that simplifies accessing a new or existing account.

Once you have an account you can create and organize a library of your preferred categories, songs and artists that can be accessed and played on your Amazon Echo Dot using voice or the app. Similar to Amazon Prime Music, there is a massive amount of content to explore on TuneIn, so enjoy the process of discovery!

Controlling TuneIn with your Echo Dot and App

If you know the radio station name or station call sign or the name of the podcast, you can use voice commands to get exactly what you want. Examples include:

- *Alexa, play The Bear, 98.1*
- *Alexa, play the Thundering Herd podcast*
- *Alexa, play A Prairie Home Companion*

If you simply request the Echo Dot to play NPR or CBS Sports, for example, the appropriate station in your market will be accessed. Ask the Amazon Echo Dot to pause or resume as desired. You can also make other requests (ask a question or add something to your to-do list, e.g.) while TuneIn is playing, and the Echo Dot will reply before returning to TuneIn.

The Echo Dot App makes using TuneIn a snap. On the left side menu of the app, select TuneIn and browse with tabs for Local Radio,

Trending, Music, Talk, Sports, News, By Location, By Language and Podcasts. Click or tap for dozens of options within each category. When you find what you want, click or tap it, and a player will begin to play the station or show.

IheartRadio: What it Offers

This network is a digital radio and content streaming service that includes thousands of live radio stations and podcasts from the US. iHeartRadio gives you the option to create personalized Custom Stations featuring your favorite artists or genre.

Music is just the beginning. Other categories include Business & Finance, Comedy, Entertainment, Food, Games & Hobbies, Health, News, Politics, Science, Spirituality and Sports. Each group offer numerous and wide-ranging options, so it will take you some time to explore what's available and find shows you want to return to often.

Setting Up your iHeartRadio Account

To enjoy iHeartRadio on your Amazon Echo Dot, no account is necessary. Simply use the Echo Dot App to explore and listen. However, with an account, you can create Custom Stations and share them with others. Select the iHeartRadio on the Amazon Echo Dot App and then click "***Link your account now***" to be taken to a special sign in/sign up page for activating you iHeartRadio with your Echo Dot.

If you don't have an account, set one up using an email address and password. You'll also be asked to provide zip code, gender and agreement with the site's Terms of Service. Another option is to link an existing Facebook or Google Plus account to create an iHeartRadio account.

Controlling iHeartRadio with your Echo Dot and App

When you click or tap the iHeartRadio tab, your options appear in the main screen. They include Search for artist or station, Browse for Favorites you've chosen while using the network.

The Browse option gives choices for Live Radio, Perfect For (Kids, Working Out, Driving and more), iHeartRadio Originals and Shows. We had a lot of fun exploring the variety of choices available.

The Originals tab features dozens of unique stations including Golden Era musical numbers, Sippy Cup for pre-schoolers, Classical Genius featuring Mozart and his contemporaries, Workout Beats, All 60s, Road Trip and Chillax.

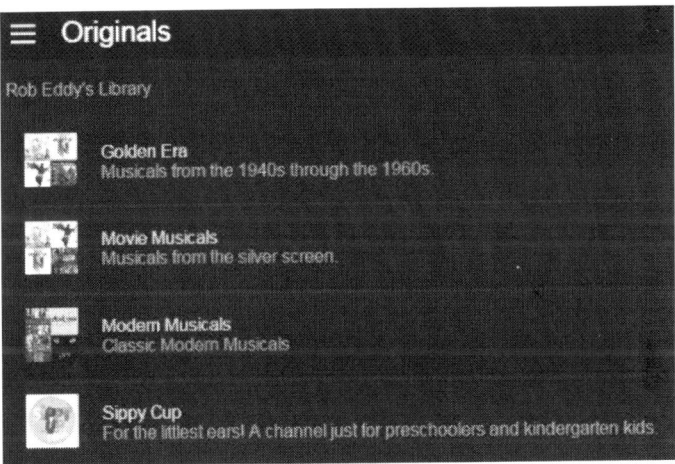

To use voice commands to play iHeartRadio, you have to know the name of the show, station or radio program you want to listen to. When requesting a radio station, your best bet to get it working is to refer to it with the exact title of the station as it is shown in the Amazon Echo Dot App.

Some use call letters; others use a name such as The Bear or Kat Country. We had to experiment in order to access some stations. For example, we requested the station listed as New Country 96.3 KSCS.

When we asked for it by "New Country," the Echo Dot retrieved a top country hits station from Amazon Prime Music. We then asked for KSCS 96.3, and the radio station was properly accessed.

You can make requests of the Echo Dot without first pausing iHeartRadio. The Amazon Echo Dot will respond and then return to playing your iHeartRadio selection. Other control options include "stop, pause, resume, continue, play." As you can see, the device responds to multiple commands for the same function.

Itunes on the Amazon Echo Dot

The popular music services iTunes does not have native support on Amazon Echo Dot. In other words, it doesn't integrate with your iTunes account in the same way as it does with the services we've already discussed.

However, if you have an iTunes account or use the iTunes app, you can still play music from iTunes using the Echo Dot. Remember, the Echo Dot is a Bluetooth speaker, so you can pair other devices like your phone, tablet or pc to the speaker and stream your iTunes music that way. The only difference is you will control your music selection with your iTunes software rather than with the Echo Dot app.

To pair a mobile device with the Echo Dot you do still need to use the Echo Dot app, select the **Settings tab** on the left and then the Bluetooth option. Choose **Pairing Mode**, and you'll be given the command to locate the Bluetooth settings on the device you wish to pair with the Amazon Echo Dot and to enter the code the Echo Dot provides.

So now go to the device you want to pair and open the Bluetooth settings there to finish the connection. That's all you need to do.

Once the device is paired, you'll choose the music that you want to play in iTunes but you will still be able to use voice commands like Play, Pause, Stop, Cancel, Resume, Continue with your Echo Dot. And remember this set up will play your music through the small Dot internal speaker. You will need a wired connection from your Dot to a external speaker if you want better sound quality.

Audible: What it Offers

Audible is an Amazon company that creates audiobooks. A standard free 30-day trial includes two free books. If you're an Amazon Prime member, the free trial is 90 days, and you receive three free books during that time. The cost of Audible after the trial period is $14.95 per month and includes one audiobook. Additional books are available in all genres and at a wide range of prices.

Follow this link for more info - www.audible.com

Setting Up your Audible Account

Choose the Audible tab on the left side of the page. It will take you to an Amazon.com page where you can sign into an existing account or establish a new account through a free trial.

If you already have an Amazon.com account, you simply sign into it and start the free trial with a couple of quick steps. If you don't have an Amazon account, starting one is as easy as providing an email and password and a few other details.

Once you've established a free-trial account on Audible, your first credit will appear on the Audible page. Browse the audiobooks and select your choice. You'll have the option of paying for it with the credit or saving your credit and using one of the forms of payment you have stored on Amazon.

Controlling Audible with your Echo Dot and App

When using the app, choose the Audible tab, and your audiobooks will appear. Select the one you want, and it will begin to play. When you pause the book to use Amazon Echo Dot to listen to music, for example, your place in the book will be marked. When you return to the book, the audio will pick up a few seconds prior to where it left off.

Using voice recognition, give these or similar command: "*Alexa...*

- *Play Audible book 'The Magic of Thinking Big'*
- *Pause the book*
- *Resume my book*
- *Go forward*

7. AMAZON ECHO DOT IN YOUR HOME

News, Weather, and Traffic

When you want a quick peek into the news of the day and current weather, Alexa is standing by with your Flash Briefing. We typically access our briefing while eating breakfast or having our morning coffee.

- To set up your personal briefing:

- Choose **Settings > Flash Briefing** on the Alexa app.

- Scroll down the list to toggle on or off news sources such as BBC, ESPN, TMZ and The Economist

- Scroll down to the News Headlines section to select from a variety of categories including Top News, U.S., World, Business, Sports and Offbeat.

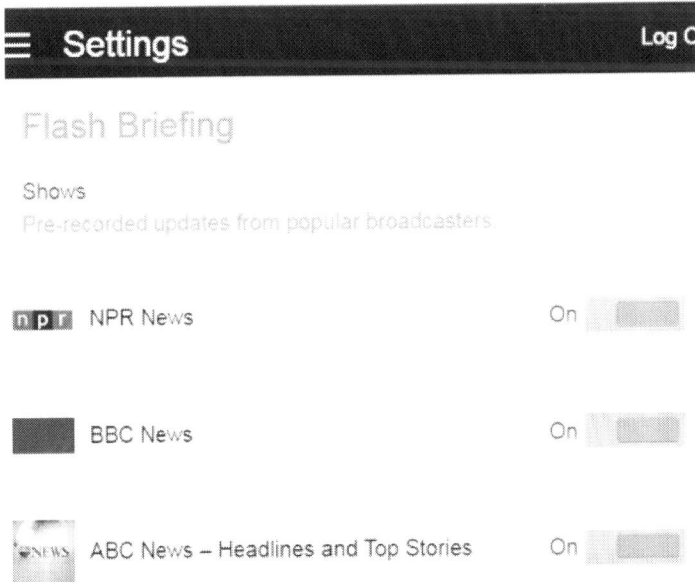

Request this feature with, "***Alexa, Flash Briefing***" or "***Alexa, what's in the news?***" The briefing may be short or long, depending on how you customize it.

When you're being briefed and want to skip to the next source or news story, say, "***Alexa, next***". You can also ask Alexa to pause the briefing or to repeat the previous story.

If the Weather toggle is on, your briefing will end with current weather conditions for the zip code you provided in the Settings for the Echo Dot device location.

The Amazon Echo Dot arrived set to NPR for the news and the local weather option turned on. Additional weather information can be accessed by asking these questions or queries similar to them. "***Alexa…***

- *What's the weather for this weekend?"*
- *What's the weather for Tuesday?"*
- *What's the extended forecast?"*
- *What's the weather for next week?"*
- *What's the weather in Chicago?"*
- *Will it rain tomorrow?"*
- *Will it rain in Los Angeles tomorrow?"*

When asked for an extended forecast or the week's weather, the Amazon Echo Dot will provide seven days of weather forecast information.

For traffic information, first go to the Traffic tab in the Alexa app settings. There, you can enter the address of the starting and ending points of your daily commute or a trip you're planning. Additional stops in between can be added and deleted using the buttons.

When you've added these details, ask a question similar to one of these: "**Alexa...**

- *How is traffic?*
- *What's my commute?*
- *What's traffic like?*

Traffic information is provided by HERE, a company owned by Nokia. The Echo Dot gives you the best route to take given current conditions including volume of traffic and construction. The expected time of your commute is given too.

Movie Times

Gone is the hassle of manually looking up showtimes online or calling the cinema. Instead, simply ask Alexa to relay information on movies showing in your town or the city you'll be heading to on a weekend getaway!

Alexa gets her movie and theater information from IMDB. Here's what you need to do once to setup the Echo Dot:

Access the Alexa App

- In the Navigation Panel, select **Settings**
- Choose your device (for example, "Jenna's Echo Dot")
- In the Device Location section, choose **Edit**
- Enter your complete address from street name to Zip code
- Choose **Save**
- If your location changes, select **Edit** to change your address

The beauty of it is that you can tailor your search with questions starting with your wake word - **Amazon/Alexa**. Here are examples of what you can do:

- Find a specific movie time: Ask, "**When is Zootopia playing?**" or "**When is the movie The Brothers Grimsby playing tomorrow?**"

- Find all your options: Ask, "**What movies are playing?**"

- Find your options for a certain time: Ask, "**What movies are playing between 8pm and 10pm?**"

- Find your options in another city: Ask, "**What movies are playing in Boston?**"

- Find movies in a specific genre: Ask, "**What comedy movies are playing?**"

- Find information about a specific movie: Ask, "**Tell me about the movie Whiskey Tango Foxtrot.**"

Find local businesses and restaurants

And Alexa isn't just good for entertainment. Once you have set up your device location you can ask about any local business. Ask for phone numbers, addresses and hours for banks, pharmacies, mechanics, restaurants...anything at all really. So for example, "***Alexa...***

...What supermarkets are nearby?"

...What are the best local restaurants?"

...Find the address for a nearby ATM"

...Find the phone number for a nearby pharmacy"

...Find the hours for a nearby bank"

By the way, the information you desire needn't be limited to your locale. If you're going on a trip you can try asking Alexa about businesses, restaurants, etc in any town or city. This infomation is provided by Yelp...so if it's on Yelp it's on your Echo Dot!

Timers and Alarms

The Amazon Echo Dot has become our go-to choice for setting a timer for baking, knowing when tile grout is ready to use and many other diverse purposes. To set the timer manually, choose **Timer** on the Amazon Alexa app, and set hours, minutes and even seconds, if you really need precision. The maximum is 24 hours. Choose **Start** to begin the countdown.

We typically use voice command for the Timer. It's as easy as saying, "*Alexa, set the timer for 45 minutes*". The timer will count down on the app, so a quick glance lets you know how much time is left. You can also ask, "*Alexa, how much time is left on the timer?*"

A series of pleasant tones sounds when time is up. The timer can be cancelled verbally with something like, "*Alexa, cancel timer*". However, the app must be used to pause or resume the timer.

You can now set multiple timers. Timers and alarms can be used simultaneously.

Alarms are equally useful, and the function can be set via the Alarm tab on the app or with a voice command like, "*Alexa, set an alarm for 6:45 a.m*". The first few times we set an alarm with the Alexa app, we forgot to toggle it On.

Keep in mind that both setting it and turning it on, just like a conventional alarm clock, are necessary when using the app. When you set an alarm using the voice recognition software, it is automatically turned on.

When the alarm goes off in its gentle tones, cancel it with the app or by saying "**Stop the Alarm**" or say, "*Alexa, snooze*" to get nine more minutes of rest. Note that the alarm must be turned on again to be set for the next day.

The timer and alarm volume is set independently in the app. Select **Settings** and **your Echo Dot**. Scroll to **Sounds** and adjust it by pressing and dragging the volume bar.

Skills

Alexa boasts a growing menu of voice-driven, real-time skills that greatly add to the device's functionality and fun. It's a good bet that these skills, introduced rather late in the evolution of the Echo Dot, will become the features that drive sales going forward.

Alexa skills provide a wealth of information about events, news, restaurants, movies and much more. There are games to entertain you and/or the kids, pithy quotes, an Alexa-led workout, a Skill for ordering event tickets and answers to the whereabouts of your car when someone else is driving it and how much gas it has. Any attempt to adequately summarize the Skills options will fall short, since this list is just the tip of the iceberg.

The majority of Skills are NOT being produced by Amazon. They're developed by the third parties (Domino's Pizza, Warner Brothers, Uber, Yelp, NBC News, Huffington Post, Capital One, to name just a few) to allow them to tap into the growing cadre of Echo Dot users. In fact, with the Alexa Skills Kit (ASK), anyone including you can build a Skill and offer it to the masses! This is the major reason for the rapid growth of Skills.

Are you ready to give Skills a try?

- Access the Alexa App

- Choose **Skills** in the Navigation Panel

- Browse the long list of Skills, and read the User Reviews of each, if you would find that helpful

- Enable any Skill you want to explore / Disable them, if desired, using the toggle

- After choosing to Enable the Skill, say something like, "***Alexa, turn on [Skill name]*** which will work in most cases, but if it doesn't, click on the Skill to learn more about setup

- Rate the Skill, if you'd like, to let other Echo Dot owners what you think of it. Choose ***Write a Review*** on the Skill detail page, and share your experience

Using Alexa Skills is a growing part of our experience, and we think you'll find them just as useful.

Just to give you a taste of what we're talking about here are two skills that are getting a fair bit of press:

Uber: There's no easier way to get a ride than this! Alexa will order you an Uber car at your request. Here are the quick steps getting started and ordering a ride:

1. Link our Uber account to Alexa

2. Add the location of your Echo Dot in the Alexa app:

- Go to Settings
- Choose the device by name, for example, Ellyn's Echo Dot
- Find the **Device Location** section, and add the address

3. Try saying *Alexa/Amazon*:

- *Ask Uber to send a ride*
- *Ask Uber to request a car*
- *Ask Uber to request an UberXL*

4. Ask for the Uber ride status or cancel the Uber

- Your Uber Services options include UberX, UberBlack, Uber-SUV, UberXL and UberSelect.

Domino's Pizza: Ordering a pizza and more for pickup or delivery from Domino's is just as easy as calling for an Uber. You can tract your order too with Domino's Tracker.

Use these quick steps to get ready, set order!

1. Set up a Domino's Pizza Profile at www.Dominos.com if you don't have one

2. You'll need a Domino's Easy Order or recent order saved within your Pizza Profile, so you might have to order online once before being able to use Alexa

3. Try saying, **Alexa/Amazon:**

- **Open Domino's**

- **Open Domino's and place my recent order**

- **Domino's, place my Easy Order**

- **Ask Domino's to track my order [once it has been placed]**

4. Type the phone number saved in your Domino's Pizza Profile into the Alexa app, for an alternate way to get tracking. - www.lyntons.com/skills.

To-Do and Shopping Lists

Like us, you might soon find these Amazon Echo Dot features indispensable. They are certainly easy to use with either voice commands or the app. Both your Shopping List and To-do List are featured on the main page of the app. Choose the one you want to manually add items or activities to.

For the To-do list, use the Add Item box on the app to populate the list. When you complete a task, select the box next to it, and a check mark will appear and the task will be lined through. Finished tasks are moved to the Completed list while the rest remain on the Active list.

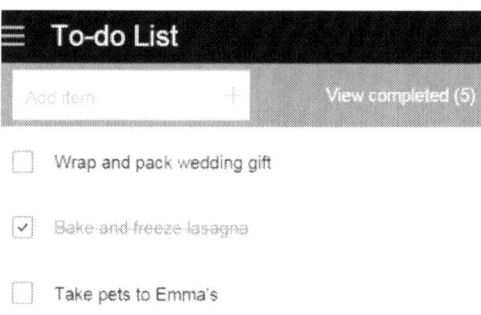

Move back and forth between these lists with the tab at the top of the page. Print the list using the tab at the top. The printer dialog box will appear for you to use to complete the process.

The Echo Dot personal assistant does a pretty good job populating the list via voice commands, though it might get a word wrong here or there. We asked, "***Alexa, add 'find CDs' to my to-do list***", and what appeared on the list was, "Find c.d.s". Then, a note to get "screen grabs" was listed as "screen grahams." Misunderstandings happen most often with uncommon words, and we doubt Voice Training would help.

There have been occasions where we've had to cast our minds back and remember what it was we asked Alexa to add because what's on the list doesn't make sense. It's inspired a blend of comedy, frustration and "ah-ha!" moments when we figure it out.

Typically, a request such as, "***Alexa, add mow the grass to the to-do list***" is clearly understood. Amazon suggests as an alternate saying something like, "Alexa, I need to organize my tools."

We've found that the Echo Dot doesn't comprehend this voice command very well. We stick to the "***Add to to-do list***" commands for best results.

Note that checking completed tasks or deleting tasks cannot be done with voice commands; the app must be used. However, you can ask the device something like "***Alexa, what is on my to-do list***" or "***Alexa, read me my to-do list***", and your request will be granted.

Using the Shopping list function is very similar to creating and managing a to-do list:

- Create it with voice commands or manually using the app
- Ask the Echo Dot to read the list to you for review
- Check off items you've purchased, and they are placed on the Completed list which can be viewed with the tab at the page's top

51

- Delete an item you no longer want

- Print the list using the tab at the top

- Items must be manually removed; voice recognition won't work for this function

The Echo Dot gives you several choices if you select the down arrow for each item. These include Search Amazon to shop for the item in a new window on Amazon.com or Search Bing to learn more about it or find other buying choices.

Sports

For sports information from your personal assistant, you've got two options. We've discussed one already; choose ESPN and/or Sports headlines as part of your Flash Briefing. The second is to get information by asking the Echo Dot specific questions such as, "***Alexa...***

- ***What is the score of the Dallas Cowboys game?"***

- ***How are the Boston Red Sox doing?"***

- ***Who won the Kentucky Wildcats game?"***

- ***When does the Atlanta Dream play next?"***

- ***What time do the Los Angeles Dodgers play tomorrow?"***

At this writing, the Amazon Echo Dot offers scores from the NFL, NBA, WNBA, MLB, MLS, NHL, PGA and NCAA Men's Basketball. If an answer isn't found, the Echo Dot usually includes a link to Bing in the Card it prepares for the query.

The Amazon Echo Dot is able to access a hit-and-miss blend of other sports information. Questions it could answer when we experimented where:

- Who won the British Open?

- Who won the Belmont Stakes?

We got, "Sorry, I didn't understand the question I heard" when asking:

Who won the NASCAR race?

Who won the Toledo Mudhens game?

So, you'll have to ask Alexa if she's a fan of your favorite obscure sport or team!

General Knowledge

The Amazon Echo Dot connects to the Cloud and its vast storehouse of data for measurements, conversions, capital cities, times around the world, mountain heights, population numbers, spelling and pronunciation, etc. Ask a question in these categories, and more than likely, you will receive an accurate answer.

The Echo Dot is excellent with trivia too. For simple answers, ask Alexa about people, dates, stars of movies or their dates of release, Academy Award winners by year and category, singers, lyricists, song and album release dates, World Series and Super Bowl winners and much more. Our guess is that you'll enjoy picking Alexa's brain to discover her fountain of interesting factoids about your favorite topics.

So, ask, "***Who starred in 'To Kill a Mockingbird'?***" and you'll get a list starting with Gregory Peck.

When you want a longer answer, for example, about Gregory Peck, say, "***Alexa, Wikipedia, Gregory Peck***". The device will read the first paragraph of the Wikipedia entry and also leave a link to the entry in the Card prepared for the query. The Card will also give you the option of searching Bing for the subject.

While the Echo Dot doesn't access recipes, it will provide a link in the prepared Card to a Bing search for recipes for pumpkin bread or whatever it is your mouth is watering for.

In the category of education, this device will be of help with spelling, definitions, conversions, simple calculations, geography, nutritional information and much more that you'll enjoy discovering.

Simon Says?

We've had a few laughs playing Simon Says with Alexa, especially when the kids got in on the act. She'll say most words, phrases and sentences except for recognized curse words. Some marginal words will get repeated. Have some fun with things like, "**Alexa, Simon says Peter Piper picked a peck of pickled peppers**", and then take it where your imagination leads!

Control Your Home

IIf you use smart home technology or plan to install some soon, then you'll be excited to know that your Amazon Echo Dot will help you maximize its convenience and benefits. Devices such as WeMo switches, Philips Hue lights and GE wireless lights are among the most popular supported by Amazon Echo Dot, but new and exciting wireless devices are constantly becoming available.

Before you can use your Echo Dot with compatible Smart Home devices you will need to have gone through the initial set up of these devices in your home.

Products vary from brand to brand but basically in order to get them working you will need to go through a process that involves downloading the manufactures companion app to one of your mobile devices and then setting up the device in your home using that app. So for example if you are planning to buy and use Philips Hue lights you will need to download the Philips Hue app from either the Amazon app store, Apple Store or Google Play store to whichever device you plan to control the lights from and then set up the lights via the app. So go ahead and get the device's companion app and set everything up making sure the device is connected to the same wi-fi network as your Echo Dot

Once you have set up the Smart Home device it's time to connect it with Alexa and your Echo Dot. From within you Alexa app navigate to **Smart Home** and click on **Get More Smart Home Skills** and search for the brand of your Smart device and when you find it click **Enable Skill.** You may well be prompted to input further account information so go ahead and do that.

> Alexa is looking for devices.
>
> Device discovery can take up to 20 seconds. If you have a Philips Hue bridge, please press the button located on the bridge and then add your devices again.

Alexa should now be connected to your Smart Home devices and accounts, so go ahead and click "**Discover Devices**" or say "**Alexa, discover devices**" and they should all show up.

Once the search is complete each device will be listed within your app. There within **Smart Home** in the app you can easily create groups of devices to be controlled with a single command. Click create group and follow the instructions, for example if you have two lights in the bedroom create a group and call it Bedroom Lights, then click on the two devices you want in that group. Once the group is created you can operate those lights with a quick "**Alexa turn on Bedroom Lights**".

You can of course control individual devices with commands like "***Alexa turn on [device name]***"

Each Smart Home manufacturer has a different Alexa skill to set up and different kinds of voice commands to give in order to use their devices. Usually this infomation can be found in the description of the Skill which you would have seen when you enabled the skill earlier. You can always go back to **Smart Home > Your Smart Home Skills** to take another look at their description.

At the bottom of this Settings page, you'll find the option to **Forget all Smart Home Settings** if you want to remove them from the Echo Dot.

It's worth saying that if you run into any problems getting Alexa to work with your Smart Home device the problem usually lies with the original set up of the home device. So do make sure your device is properly set up and working as per the manufacturers guidelines before trying to connect it to Alexa. Again, make sure your devices are connected to the same wifi network as your Echo Dot.

Shopping

For now, your lone shopping option is to re-order products from Amazon that are Prime eligible. For example, we recently ordered wool socks before traveling to the Black Hills of South Dakota to hike.

Alexa was given the command to "**Order socks**". The device located the previous order and confirmed the brand name and number of pairs in the package. It then asked us, "Should I order it?" An "**Alexa, yes**" was given, and the order was placed.

That reminds us: We've got into the habit of always saying the word "Alexa" every time we talk to our Echo Dot, but when Alexa asks you a question to clarify a request you've made, you don't have to include "Alexa" in your answer. Amazon Echo Dot is already listening for your answer.

There are a few things to do on the Alexa app to manage your purchases. Go to **Settings > Voice Purchasing**. The Echo Dot comes with **Purchase by Voice option** in the ON position. It can be toggled to OFF, if you prefer.

Next, if you're concerned about unauthorized purchasing, you have the option of requiring a confirmation code. In the app, type in any 4-digit code you want. The next time you place an order, Alexa will say, for example, *"tell me Jenna's voice code,"* which must be given prior to confirming that you want the order to be placed.

Finally, from the app, there's a link to your 1-click payment methods on Amazon.com where you can add, delete or edit credit or debit card information, make address changes and manage related details.

Amazon Echo Dot & Google Calendar

Keeping track of our schedules on a calendar in the cloud gives us the chance to check it from anywhere and to find out what each other has going on. We've used Google Calendar for more than a year and highly recommend it.

It offers calendar views for Day, Week, Month and 4 days. You can also create an Agenda that automatically shows holidays and the birthdays of people you're connected to on Google Plus. Items from the Agenda can be copied to your calendar with notes.

We're telling you all about Google Calendars because they can be used in conjunction with your Echo Dot. So let's look at setting up a Google Calendar, if you don't already have one.

Setting up Your Google Calendar Account

Sign into an existing Google account by selecting your profile picture (it will be generic blue if you haven't added one) on any of the Google pages you use (Search, Gmail, Google+, etc.).

Create a new account on any Google page. You'll either find a *Sign Up tab* or a *Sign In tab* that will take you to a page where you can create an account. It takes a minute or so. Through your one Google account, you will be able to set up and use any Google service.

Here's how to begin using Google Calendar once you have a Google account:

- Select ***My Account***

- Locate Google Calendar among your options, and choose it

- On Amazon Echo Dot, open the Amazon Alexa App

- Choose ***Settings*** > ***Calendar***

- Click "Link the Google Calendar account"

- If you have more than one Google account (such as personal and business), select the one you want to link

www.google.com/calendar

When we first tried it, we got an error message that opened in a new window. We simply closed the window and tried again, and got a window with a message that "Alexa would like to: View your email address, View your basic profile info and Manage your calendars".

Select the circled "i" next to each topic for more information, or simple choose "***Accept***". This step failed the first time too, but succeeded on the second attempt.

Accessing Your Google Calendar With the Alexa App

Once you have connected Alexa to your Google Calendar you can check and add events with just your voice. Ask Alexa to add something to your calendar and she will, like a helpful personal assistant, ask you questions "For when?" "For what time?" "What's the name of the event?". Of course you can just give Alexa all this information at once by saying "Alexa, add Dinner With Friends to my calendar for December 8th at 8pm" and Alexa will repeat the information back to you and add it to your calendar...great!

And of course it's just as easy to check your calendar with requests like "***Alexa, when is my next event?***" or "***What's on my calendar?***"

Add an Adult to Your Echo Dot Household

When Alexa joins the family, don't be surprised if your significant other wants to be BFF's with her! Each Amazon Echo Dot Household account supports up to two adults.

Our favorite perk of being on the Household account together is that we are able to share our music and content libraries with one another. For example, at the top of the Amazon Prime Homepage on the Alexa app, the top tab allows us to toggle between our two Libraries.At this writing, Amazon Echo Dot cannot access children's content that may be part of your Amazon Household account.

Additionally, we can both use Voice Purchasing which is discussed in the Shopping section of this guide. Voice Purchasing can be set up under **Settings > Voice Purchasing** where you'll select your options.

To add a household member, go to the **Settings** and select **Household Profile**. A new tab on your browser will open where you'll sign into your Amazon account and be taken to a page with the heading "Invite a household member." The page includes a brief description of the benefits.

When you continue, the screen will ask you to pass the device you're using to the person you're inviting to join the Household account. They'll sign into their Amazon account.

In a minute or two the second user's music and content libraries will be available for sharing. To leave the Echo Dot Household or remove a member, go to **Settings > In an Amazon household with...** There, select your choice.

As a precaution, a popup window appears asking you to confirm or cancel your selection.

Voice Commands and Things to Try

We hope that we have helped you find your way around your Amazon Echo Dot and you now have a good working knowledge of the features available and how to use voice commands. Our aim has been to go beyond a dry repetition of the basic functions and share with you how we actually use the Echo Dot in our daily lives.

We of course appreciate that you will probably have your own preferences so before we move on we would like to draw your attention to a very useful reference.

There is a full and extensive list of voice commands that you can find in Amazon's help pages online.

This information is repeated within the Alexa app, just scroll down the left column and click on "Things to Try". As well as voice commands you can also keep up to date with new features.

8. THE ECHO DOT AND IFTHISTHENTHAT (IFTTT)

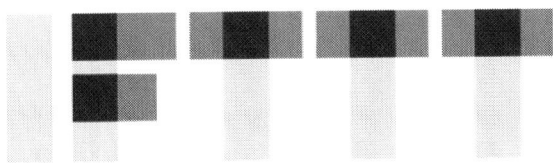

Have you ever lost your phone and asked someone to call it, so you could hear it ring and locate it? Now with the help of something called **IFTTT**, your Echo Dot can be used for dozens of tasks just like this that it cannot accomplish by itself.

IFTTT is a third-party service supported by the Amazon Echo Dot. It creates rules, called Applets, which allow your devices to work with other apps and websites to accomplish tasks. Triggering a call to your phone by, for example, asking Alexa what's on your to-do list is just one of many ways you can expand the functionality of your Amazon Echo Dot with IFTTT.

The service sounded like little more than a novelty to us when we first explored it, we must admit. However, we were soon enthusiastic about the connection between the Echo Dot and IFTTT and use it on a regular basis, if not yet daily.

The letters IFTTT stand for IF This Then That. Let's explain using the example we've mentioned:

IF This: You ask "Alexa what's on my to-do list?"

Then That: Alexa places a call to your cell phone.

Combinations like this are called **If Applets**, If I do This, Then That will happen. Quite a few of them work with Alexa. The "If" side of the equation is known as the Trigger; it is the event that causes the action to occur.

The fact that a trigger such as asking Alexa to review your to-do list seemingly has nothing to do with a phone call being made shouldn't confuse you. You're simply asking the Amazon Echo Dot to do something it knows how to do and then connecting technology to that function that will cause something unrelated to occur.

Still confused? Read on...

Start by Activating the Amazon Alexa Service on IFTTT - Getting started took us about three minutes. Here's how to do it:

Go to the Amazon Alexa Service on IFTTT (https://ifttt.com/amazon_alexa), and Sign In to an existing account or Sign Up for your first account

- Sign in
- Select Connect
- Sign into your Amazon account in the pop-up window

Using Your Echo Dot With IFTTT

Once we were signed in and set up we set aside a half hour or so to familiarize ourselves with the Applets (https://ifttt.com/amazon_alexa) that can be accomplished by creating relationships between the Echo Dot, apps and other devices.

A few hours later, we were still having fun exploring and trying many of the applets. At this writing, there are over 250 applets on the IFTTT Alexa page, and more are being written regularly.

Here's a tip: Download and explore the IFTTT app using your Android or iOS mobile tablet or phone rather than a PC or MAC. You'll likely want to use some of these Applets, and to do so, you'll find that some of them require you to download the other necessary 3rd party apps to your phone or tablet. It is these 3rd party apps that actually implement the action that is triggered when you give a specific order to Alexa.

Top Applets for Alexa

Our "top" Applets might not be yours once you've had a chance to explore the possibilities, but we have found these to be very handy. You'll see that each one of these has an "if" task that will create a "then" result.

Ask Alexa to locate your phone: (www.lyntons.com/findphone) Let's see how the example from above works in practice. Even if you don't activate this applet now, it's a great one to keep in mind for that inevitable time when the phone goes missing in the couch.

- **If:** You ask, "*Alexa, trigger find my phone*"

- **Then:** Alexa will call your phone

- **How:** To set up this applet click the ***Turn on*** tab to connect to the Phone Call Channel. A new window opens, send your phone number to the Phone Call Channel and you'll receive a phone call with an activation PIN. Enter the PIN and you're done. Now when you say the trigger "*Alexa, what's on my to-do list?*" your phone will receive a call.

Find My Phone: (www.lyntons.com/find) is a similar related applet that turns up your android phone's volume to make it easier to hear and locate when it is called.

Email yourself your shopping list via Echo Dot: (www.lyntons.com/list) Take your shopping list to the store with you on your phone.

- **If:** You ask, "*Alexa, what's on my shopping list?*"

- **Then:** The Echo Dot will email you the entire list.

- **How:** First, add your email address to the box on the ITFFF applet page. Then, make your request of Alexa, and the email of your shopping list will show up in your inbox shortly.

If you already use OneNote, send your shopping list there in short order with - www.lyntons.com/note.

You can email yourself your to-do list just as easily - www.lyntons.com/todo.

Or, transfer your to-do list - www.lyntons.com/evernote - or your shopping list - www.lyntons.com/evernote1 - from the Alexa App to Evernote.

In addition to sending a short email or emailing yourself your to-do or shopping list, there are other email-related If applets you might find helpful:

- Accomplish something from your to-do list and send a clapping gif - www.lyntons.com/clap - to friends or family. Yes, okay a bit silly but surprisingly satisfying.

- Send yourself the new items you've added to your to-do list - www.lyntons.com/newtodo

- Send your to-do list to a Wunderlist account - www.lyntons.com/wunder - or to OmniFocus - www.lyntons.com/omni

- Most of the applets involving email allow you to include up to five email addresses.

Print your shopping list with an HP wireless printer: (www.lyntons.com/print) Use voice to create a list and a simple command to have the list printed. Oh course you will need an HP wireless printer and an HP Connected Account.

- **If:** You ask, "*Alexa, what's on my shopping list?*"

- **Then:** Your list will print.

- **How:** The Echo Dot will send the list to the printer when you connect via the applet page.

Philips Wi-Fi lighting is hugely popular with smart home enthusiasts, and the number of Applets that make use of Philips hue bears that out. Here are a few of the many other applets to try with Philips hue lights:

- Ask Alexa to change the light color (www.lyntons.com/colorswap)

- Ask Alexa to toggle all hue lights on or off (www.lyntons.com/togglehue)

- Change light color when a set timer goes off (www.lyntons.com/timer)

- Change hue light color each time a new song plays (www.lyntons.com/songchange)

- Blink hue lights when a set timer goes off (www.lyntons.com/blink-hue)

Turn on a WeMo switch when an alarm goes off: (www.lyntons.com/wemo) WeMo switches are Wi-Fi connected switches. They are plugged into standard outlets. Then, a lamp can be plugged into the WeMo and turned on. From there, the WeMo switch can be toggled on or off with a Wi-Fi device and app.

- **If:** An Alexa app alarm goes off.

- **Then:** The WeMo switch will turn on.

- **How:** Add the WeMo switch to your Alexa app at ***Settings>Smart Home***. Manually turn the switch on. On the applet page, connect to the WeMo Switch channel. Then, set an Echo Dot alarm for when you want that switch to be energized.

Set the thermostat temperature: (www.lyntons.com/temp-nest) This applet is for the Nest wireless thermostat, but a applet is available for the popular Ecobee thermostat (www.lyntons.com/temp-ecob) too.

- **If:** You add an item to your to-do list.

- **Then:** The applet will allow you to set the thermostat to the desired temperature.

- **How:** On the applet page, connect to the Nest Thermostat Channel, and follow directions from there.

You've Got Options

As you browse the applets, you'll find more than one for some of the same tasks. This simply means that the applets were created by different users. In our experience, any two applets that do the same thing work equally well, but we haven't tried them all. You might have to experiment with a couple of Applets to find the one that works best for you.

The My Applets Page

Once you use an Applet, it will be added to your **My Applets** page to access again later. Simply select any recipe from that page to use it as you created it. You can also use the Edit button, discussed below, to tweak it for each use.

On each applet are your options:

- **The On/Off Button:** This allows you to turn the applet off when you don't want it to respond to trigger input and back on again when you do. For example, we keep the applet "Find your phone by asking 'what's on my to-do list'" off. It's ready to switch on should we misplace the phone, which happens more than we care to admit!

- **The Reload/Check Now Button:** Selecting this button

checks for trigger data immediately. At this writing, IFTTT checks for triggers every 15 minutes for most applets. If you want to trigger the applet immediately, select this button.

- **The Edit Button:** Selecting this button takes you to an edit page where you can turn the applet on or off, check it, view the log and delete the applet from your My Applets list. You can also publish applets you've created. For creating applets, see below. On the Edit page, you can also change the email recipient or add recipients, change the text on the subject line of the email that goes out and include a URL as an attachment to the email. If this sounds confusing, it will become clear with just a little experience managing applets, as it did for us.

Note: When you've made edits to any applet, be sure to select the *Update* button to put them into effect.

Create Your Own Applets

Now, if you've been paying attention you will have noticed that many of the applets have the same triggers, for example many of the applets are activated when you say "*Alexa, what's on my to-do list*".

The reason for this is that currently there are only a limited amount of trigger commands available; although we suspect this will change in the future.

So what do you do if there are two applets you want to use but they both have the same trigger command, after all you don't want to say one command and have two actions happening at the same time? The answer is to create your own new applet. Near the top of the **My Applets** page, you'll notice the **New Applet** button. Select it to be taken to the Applet Maker.

The steps are simple to follow, but here's an overview:

- **Step 1: Choose a Service**

For the Echo Dot, IFTTT calls the trigger service Amazon Alexa so scroll down or search for it in the search box. It's the only trigger that works with the Echo Dot. If you're creating applets for other devices, you'll find channels for Android and iOS devices and dozens of apps for services, retailers and more.

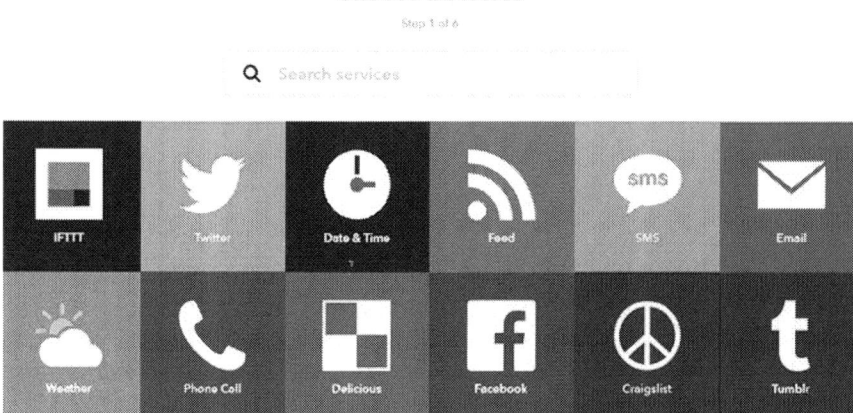

- **Step 2: Choose a Trigger**

Here, you will select the specific "If This" action you want as the trigger. Options include asking Alexa a question such as what's on your to-do or shopping list, adding an item to either of those lists or an Echo Dot timer you have set going off.

The most helpful trigger for us is Say a specific phrase which basically allows you to set up any action and trigger it with "Alexa trigger… (a phrase which you decide on)"…so just like the very first applet we discussed you can say "Alexa trigger find my phone" and your phone will ring, in this example the phrase you decided on is "find my phone".

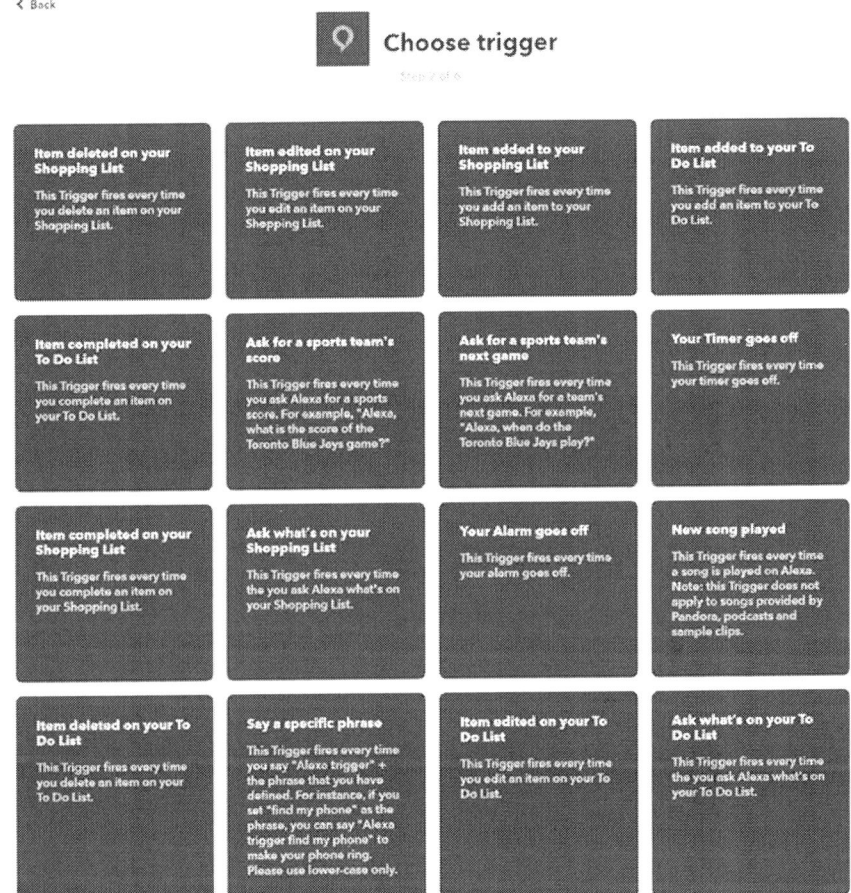

Once you've create your trigger you've completed the "If This" half of the applet. You should now see a screen showing "If [Echo Dot symbol] then that" appears with the "that" in blue for you to select.

Its purpose is to show you that you've completed the "If This" steps, but in our opinion, seems to be a useless step.

- **Step 3: Choose an Action Service**

Here, you select the device or app that you want to carry out the

"Then That" action such as send an email, set a thermostat, make a phone call, load a photo to Facebook or Pinterest, adjust lighting or turn on an appliance.

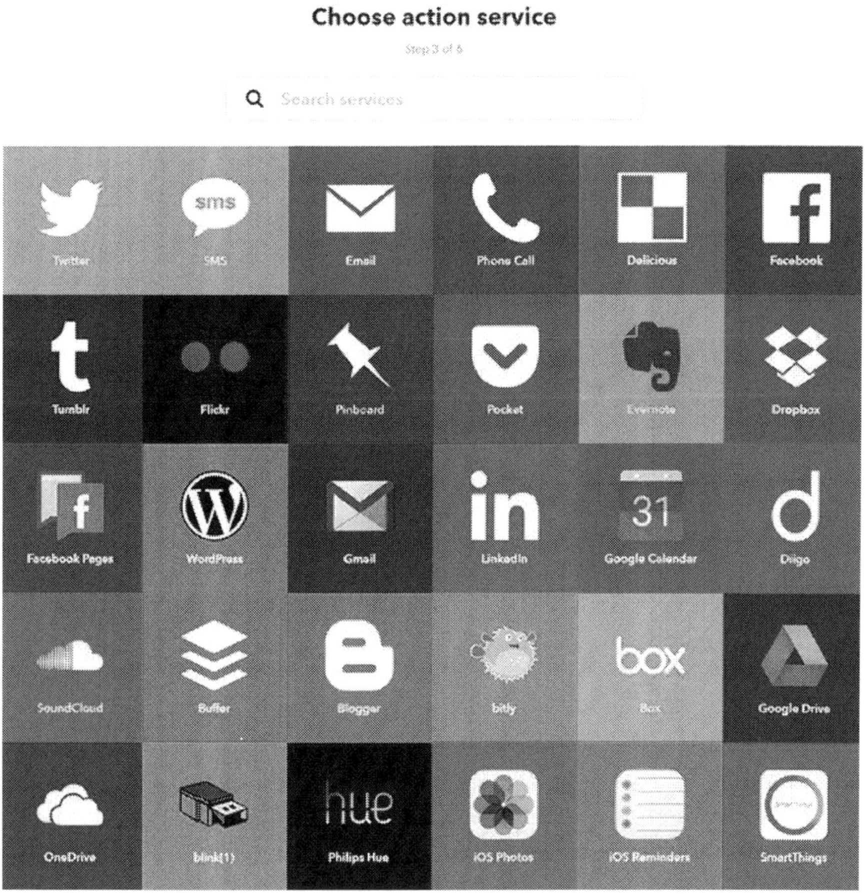

- **Step 4: Choose an Action Channel**

You've selected a "service" to carry out the "Then That" action, now specify just what you want to happen. For example, if you've selected your Philips hue lighting, you can turn it on or off, cause it to blink or change colors or perform any of its other capabilities.

- **Step 5: Complete Action Fields**

In this step, additional specifics, the "ingredients" of your applet, are added. For example, if the Channel (Step 4) is Gmail and the Action

(Step 5) is to send an email, in Step 6 you'll choose the email's recipient, subject and text. If the Channel is Nest and the Action is to set your thermostat, Step 6 will be to select your preferred temperature and other options such as choosing the heating or cooling function to complete the desired Action.

M Complete Action Fields

Send an email

M To address

M Subject

Score of the AskedTeamName Game

- **Step 6: Create and Connect**

You'll see the details of the applet in the Applet Title box. If it is correct, choose the Create Applet button, and your applet will be created. You'll be taken to your My Applets page, and the newly created applet will appear at the top of the list.

We've tried a couple dozen existing applets and created a handful more. We're still exploring and adding applets, and we've found that several have become very useful for adding things to our to-do list or calendar, creating reminders and managing parts of our home automation system.

As with any new gadget or app, there's a short learning curve for IFTTT basics and then an ongoing period spent finding, creating and customizing applets to give us maximum convenience and functionality. We're enjoying the process, and think that you will too.

9. TROUBLESHOOTING YOUR AMAZON ECHO DOT

Our experience with the Amazon Echo Dot has been pleasantly trouble-free, and we expect the same for you. However, some users do experience problems. The vast majority of them are minor and can be easily solved.

Below is a list of issues and the steps you should take to try to resolve them. We reference the Things to Try (www.lyntons.com/try) list several times. That list can be accessed on Amazon, and it is also included in your Alexa App.

The Echo Dot Doesn't Understand You

Try these:

- Eliminate background talking or other noise that could be confusing the voice recognition technology

- Speak clearly and more slowly

- Complete Voice Training with the Echo Dot, so that it improves its ability to understand you. Visit the Alexa app's Voice Training section.

- Phrase your request differently; The list of Things to Try includes topic pages where you'll find requests the Echo Dot understands.

The Echo Dot Can't Give You an Answer

Try these:

- Repeat the question

- Rephrase the question; See the list of Things to Try for help

- State the question more specifically or more broadly

- Let the Echo Dot know you're asking a question by saying, for example, "*Alexa, question: What's the capital of Romania?*"

- Keep in mind that the Echo Dot can't answer every question, though its capabilities continue to expand. By exploring the Things to Try list, you'll gain a better understanding of what questions the Amazon Echo Dot can and cannot answer.

The Echo Dot Plays the Wrong Music

Try these:

- Check your Cards (your history) on the Alexa App Homepage to see what the app heard you say

- Check to see if the music you requested is in your library or available from Amazon Prime Music

Voice Purchasing Code Doesn't Work

Try these:

- Locate the On/Off toggle on the Echo Dot App under **Settings > Voice Purchasing** and make sure it is on

- In the same location, **Settings > Voice Purchasing**, make sure that you've chosen a code and are using the right one in your attempt to make a purchase

Amazon Echo Dot Remote Won't Pair

Try these:

- Make sure that the two AAA batteries are fresh

- Select **Settings > Pair Remote** in the Alexa app, and then hold the ***Play/Pause*** button on the remote for five seconds while the Echo Dot searches for it

- Check the light on the top of the Echo Dot, and if its spinning and purple, more than one remote was located

- If two remotes were located, repeat the process and press ***Play/Pause*** on the remote you prefer to connect with

- If you know that Echo Dot is paired with a remote you don't want to use, go to ***Settings > Forget Remote*** to clear it before pairing it with the preferred remote

- **Note:** Only One Amazon Echo Dot Remote at a time can be paired with the Echo Dot

Echo Dot Won't Connect To Wi-Fi

Try these:

- If not connected, go to the Alexa app Homepage, open the navigation panel on the left, and select ***Settings > Echo Dot > Set up a New Echo Dot***

- Press and hold the ***Echo Dot Action*** Button for about five seconds, and the light ring will turn orange while your mobile device, if using one with the Echo Dot, connects to it

- Choose your Wi-Fi network from the list that appears, and enter the network password if necessary

- Select ***Connect***

- If your network isn't on the list, scroll down to select ***Re-scan*** to search again or choose ***Add a Network*** and follow the instructions

- When connection to the Wi-Fi network is completed, a confirmation message will appear in the app

Echo Dot Won't Connect to Bluetooth

Try these:

- Bring your Bluetooth phone or other device within 30 feet of the Amazon Echo Dot

- Turn on the device's Bluetooth connection

- Make sure the device is paired with the Echo Dot, and if it isn't, say, "*Alexa, pair my phone*"

- Check the Card on your Alexa app to see if it is hearing, "Connect".

- Alternatively, find the Echo Dot in your device's Bluetooth settings menu, and manually select the Amazon Echo Dot to attempt to connect

- If you're still having trouble, go to S*ettings* > *[Your Echo Dot's name]* > *Bluetooth*, and select *Remove* to clear all paired devices

- Pair your Echo Dot to your Bluetooth device again

- Unplug the Amazon Echo Dot's power cord for three seconds, and connect it again

- Pair Echo Dot to your Bluetooth device (www.lyntons.com/pair)

Echo Dot Can't Discover a Smart Home Device

Try these basics first:

- Make sure the device is on the list of Supported Home Devices (www.lyntons.com/supported) for Amazon Echo Dot

- Set up the manufacturer's companion app for the device, if you haven't done so

- For Wink and Wink-connected devices, link your Wink account by following the steps at Connect a Hub Service (www.lyntons.com/connect-a-hub) to Amazon Echo Dot

- Use the device's companion app to check to see that it has been properly set up, and if it isn't, refer to the device's guide or website

- Use the devices companion app to download any available updates including fixes to ensure the best connectivity

- Connect your Amazon Echo Dot to the same Wi-Fi network as the device, so Echo Dot can discover it

- To do this, go to the Alexa App and update your Wi-Fi network at **Settings > [Your Echo Dot] > Update Wi-Fi**; we suggest you use your home network since work or school networks might not allow unrecognized devices to connect

- If there is still no connection, say, "**Alexa, discover my devices**"

- If the device is a Philips hue bridge, press the button on the bridge and hold it while instructing the Echo Dot to discover your devices

- If the Echo Dot says, "Discovery is complete. I couldn't find any devices," check again to make sure the Echo Dot and the device are connected to the same Wi-Fi network

Note that some devices can only connect to a 2.4GHz Wi-Fi network. These include:

- WeMo Insight Light Switch
- WeMo Insight Switch
- WeMo Switch
- Wink Hub

5 GHz networks: If you're using your Echo Dot on a 5GHz Wi-Fi network, switch to a 2.4GHz network to discover and connect the device. If your router is dual band, and you don't know the name of the 2.4GHz network, check the router settings on your computer or contact the router's manufacturer for help.

Enable SSDP or UPnP on your router: This can be done via the router settings on your computer. If you can't locate them, contact the router manufacturer or its website.

Choose an easy name: If your Smart Home devices are discovered, but Echo Dot doesn't process the request, perhaps it is because you've assigned a difficult name to the group.

Change it to something simple like *Kitchen Lights* or *Hall Switch*. If the device or group was named via the manufacturer's companion app, that's where the name change must be made.

Restart your Amazon Echo Dot by unplugging it and plugging it back in. Restart your Smart Home device by following the user guide for it.

Reset your Amazon Echo Dot: Occasionally a seemingly stubborn problem is solved easily by resetting the Amazon Echo Dot. First, simply restart it as described immediately above. If the issue isn't fixed, then reset the Echo Dot using the reset button on the bottom of the unit.

Reset:

- Use paper clip or similar tool to press and hold the reset button for five seconds

- Watch the top light ring as it turns orange and then blue

- When it turns off and then blue again, this indicates it is in setup mode

- Open the Amazon Alexa App to connect your device to a Wi-Fi network and register it to your Amazon account

If you need to set-up your device again at any time, you can return to **Chapter 2**

10. SECURITY

In this era of breaches of financial data and government agencies monitoring private conversations, it's no surprise that Amazon Echo Dot users are concerned about privacy issues. When we read that the Echo Dot processes and stores information in the Cloud, it sounds like a large, nebulous location where it might be difficult to maintain security.

The Cloud isn't a huge pool that anyone can access. Instead, it is a series of large computer systems and servers owned by private and publicly traded companies and located around the world. These systems support the Internet and its millions of sites and billions of devices.

Cloud computing is growing steadily, and you might already do business with companies using the Cloud. These include Apple, Google, Netflix, Flickr, Yahoo Mail and Microsoft. Banks are using the Cloud for storing and processing date and a rate that is increasing. According to recent analysis, the average bank uses more than 800 Cloud services.

The computers and servers that make up the Cloud use the best security available, and there's a reasonable expectation that the information you speak to your Amazon Echo Dot will be secure. Besides, most of us aren't making requests of Alexa to remember our social security number or credit card number. Most requests we make would be of no value to anyone or risk to us should they become known.

Okay, Let's Cover Some Specifics About the Security of Your Amazon Echo Dot Account.

What you say to the Echo Dot is processed and stored in the Cloud: By now, you're probably familiar with the Cards the Amazon Alexa app produces each time you give it a command or ask a question. They're available for review on the app homepage. Each card represents a voice recording that is also stored in the Cloud. Here is more information straight from the Amazon Echo Dot FAQs:

"How do I know when Amazon Echo Dot is streaming my voice to the Cloud?"

When Amazon Echo Dot detects the wake word, when you press the action button on top of Amazon Echo Dot, or when you press and hold your Amazon Echo Dot remote's talk button (sold separately), the light ring around the top of your Amazon Echo Dot turns blue, to indicate that Amazon Echo Dot is streaming audio to the Cloud to process your question or request. When you use the wake word to talk to Amazon Echo Dot, the audio stream includes a fraction of a second of audio before the wake word, and closes once Amazon Echo Dot has processed your question or request."

The Echo Dot uses the recordings, especially when you give feedback about whether or not it heard you correctly, to improve its voice recognition abilities. Over time, it is learning to understand you better in order to give you better results.

You can delete the cards, and the voice recordings are deleted too. Here is what Amazon says regarding this practice, again from the FAQs:

"How do I delete individual voice recordings?"

You can delete specific voice interactions with Amazon Echo Dot by going to **History** in Settings in the Amazon Alexa app, drilling down for a specific entry, and then tapping the **Delete** button."

"Can I delete all my voice recordings?"

Yes, you can delete the Amazon Echo Dot voice recordings. Doing so will delete related Home Screen cards, and may degrade your experience using Amazon Echo Dot. To delete the recordings associated with your account, visit **Manage Your Content and Devices** at www.amazon.com/myx and select **Amazon Echo Dot**, or contact customer service.

While a deletion request is being processed, the Amazon Alexa app may still display and allow you to play back the voice recordings that are being deleted. You can delete specific voice interactions with Amazon Echo Dot by going to **History** in Settings in the Amazon Alexa app, drilling down for a specific entry, and then tapping the **Delete** button."

In conclusion, the information you speak to the Amazon Echo Dot is very likely as safe as your banking and credit card information. If you have any remaining concerns, be selective about what questions and commands you give to the Echo Dot, and delete any you would be uncomfortable about someone else accessing.

11. YOUR FUTURE WITH ALEXA

We expect the practical benefits of using our Echo Dot to go through the roof in the days ahead. Just as an example, when our Echo Dot arrived a few months ago, before we published this book, it didn't support Google Calendar.

However, within a month, that changed, and we happily linked our Amazon Echo Dot to Google Calendar. Just recently, support for the Pandora music service was added, and while that wasn't huge for us, a couple of nieces who visit often thought it "way cool." And just before we went to print the Skills section appeared, so clearly there's a lot more to come.

So What Lies Ahead? Let's Make Some Educated Guesses

Amazon will develop and add new services and form more partnerships with third-party providers. This is happening consistently. Since we've fired up our Echo Dot, ESPN Radio has joined Pandora and Google Calendar as new additions.

We expect Amazon to sell an enormous number of Echo Dots, and that will create growing demand for services providing customizable news, music, sports, weather and specialty channels. As partnerships are formed, your Echo Dot will be updated to take advantage of the new options.

IFTTT is hot, and so are the technologies such as home automation and Cloud computing that make many of the recipes possible. In addition, most of us will purchase additional Wi-Fi gadgets in the coming years, part of the burgeoning "Internet of things" (devices connected and controlled online).

These trends and the potential of IFTTT mean that your Amazon Echo Dot will likely become far more useful for important parts of your lifestyle such as:

- Home automation of lighting, HVAC, appliances, home theater, sound system, etc.

- Scheduling and time management

- Shopping

- Social media

- Control of your smart devices

Exciting new apps for the Amazon Echo Dot are on the way. We anticipate this because Amazon has invited a chosen array of tech-savvy developers to try their hand at creating apps in addition to the standard Alexa app that supports its functionality.

The word is that the software development kit the developers are given by Amazon supports Java, JavaScript, Node.js, Python, PHP and Ruby programming languages. This means that all of us can expect diverse and highly functional apps to be developed that will make our experience with the Amazon Echo Dot that much more awesome!

A Final Quick Reminder About Updates

As we mentioned at the start of this book, The Amazon Echo Dot and indeed all media streaming services, like Apple TV, Roku and the Chromecast, are still in their infancy. The landscape is changing all the time with new services, apps and media suppliers appearing daily.

Staying on top of new developments is our job and if you sign up to our free monthly newsletter we will keep you abreast of news, tips and tricks for all your streaming media equipment.

If you want to take advantage of this, sign up for the updates here (www.lyntons.com/updates): Don't worry; we hate spam as much as you do so we will never share your details with anyone.

Made in the USA
Lexington, KY
12 January 2017